Ethics in the Doing of Technoscience

"From Titan's deep-sea implosion to the overuse of diamorphine in an elderly care unit, Hayes and Maslen's edited collection dissects compelling stories of failure and wrongdoing. With the appeal of a true crime podcast, the authors take us close to the action, revealing the messy reality of ethical decision making in technoscientific settings."

—Rebecca Olson, Professor of Sociology, University of Queensland, Australia

"Starting from real situations in very diverse domains and environments, this book unveils the complexity of professional ethics: its social nature, its variability across contexts, its contestation. In an original move, the authors get beyond ethical principles and the organisational scale of analysis to examine ethics in the making from the perspective of professionals on the ground. This is a rare and welcome contribution."

—Corinne Bieder, Scientific Director, Foncsi (Foundation for an Industrial Safety Culture), France

Jan Hayes • Sarah Maslen
Editors

Ethics in the Doing of Technoscience

Navigating Professional Frameworks and Organizational Muteness

palgrave
macmillan

Editors
Jan Hayes
RMIT University
Melbourne, VIC, Australia

Sarah Maslen
RMIT University
Melbourne, VIC, Australia

ISBN 978-3-032-20001-3 ISBN 978-3-032-20002-0 (eBook)
https://doi.org/10.1007/978-3-032-20002-0

Cover illustration: No credit

This Palgrave Macmillan imprint is published by the registered company Springer Nature Switzerland AG.
The registered company address is: Gewerbestrasse 11, 6330 Cham, Switzerland

Preface

Until the late twentieth century, when things went wrong with technology it was usually individuals close to the sharp end who were blamed. In recent decades, research into causation of major industrial accidents and failures in healthcare redirected attention away from frontline workers such as doctors and airline pilots to systems and organizations with a burgeoning of research into culture as the driver of professional behavior. While we welcome moves to look beyond blaming frontline workers, this approach ignores the richness and complexity of professional practice and the important ethical dimension of high stakes decision making in complex sociotechnical systems.

After more than a decade of working with engineers in the gas pipeline sector, we know that there is a lot that we can learn by focusing attention on the interactions between professionals, and between professionals and the public when it comes to high stakes decision making. Engineers speak about how their experiences have "blooded" them, influencing the strength with which they advocate for certain courses of action. They are also acutely aware of the dance that they must do to navigate relationships and organizational processes, which can involve remaining mute on ethical matters—at least on the surface. Looking closely at professional practice also shows how, even in archetypally technical work surrounded by standards and regulation, decisions are rarely simple. We tend to think that the distinction between right and wrong or ethical and unethical decisions is clear cut, but in the context of professional decision making ethical ambiguity can be high.

Our long-standing interest in this topic motivated us to arrange a panel on applied professional ethics in the doing of technoscience at the Society for Social Studies of Science (4S) conference in Honolulu, USA, in November 2023. We were overwhelmed by interest and ran three sessions on the topic. We also met scholars at this conference beyond our sessions who were grappling with professional ethics. A core group of contributors in this volume came from these connections made at 4S. We ran another panel on professional practice at the International Sociological Association Forum in Rabat, Morocco, in June 2025. Another author joined our project through this pathway. Other contributors came to the project via our academic networks.

In leading these conversations, we have made a conscious decision to move beyond studies of ethics frameworks focusing instead on ethics in practice. By focusing on the micro scale of analysis, we capture the negotiated nature of decision making in the face of ethical ambiguity characterized by differing values and priorities. Our goal has been to produce a book for social scientists with an interest in science and technology studies (STS), professional practice, decision making, or risk management. How technoscientific work is done is foundational in these fields although links to ethics in practice are thin. We also welcome readers coming to this volume as practitioners in technoscientific fields. We have found that professionals often seek out the stories of others in their own domains and in adjacent fields for windows into how they meet the challenges that they face. Many of the chapters in this volume are not only discussions of scholarly significance. They are also stories of things going right or wrong in given circumstances, and professionals can take lessons from them.

As editors, we come to this work with quite different backgrounds. Jan moved to academia and to sociology after a 20-year career as an engineer working in industry. Sarah came to expert decision making through an academic pathway, finding cases of practice in medicine, engineering, and most recently synthetic biology viewed through a sociological lens most compelling. Our interdisciplinarity and ability to speak the language of different sectors is a strength of our joint work, bringing intellectual rigor and a strong commitment to understanding and supporting professionals on the ground. Our contributors are sociologists, STS and engineering studies scholars, safety scholars, and legal scholars. They each have their own stories of how they came to their interest in professional ethics, but we have intellectual commitments in common. We owe thanks to our chapter authors for engaging with ethics in practice in such diverse and interesting ways.

Finally, we thank Zhe Zhang and Barbara Cox for their editorial assistance. We also thank the team at Palgrave, including Rachael Ballard who was the commissioning editor for the project and Naveen Dass for her guidance through the publication process.

Melbourne, VIC, Australia — Jan Hayes

Canberra, ACT, Australia — Sarah Maslen

Contents

1 **Introducing Professional Ethics in Practice** 1
Jan Hayes and Sarah Maslen

2 **"Keep as 'VERY CONFIDENTIAL!!!!'": How the Behavior of Technical Professionals Contributed to the Grenfell Tower Disaster** 27
Jan Hayes, Orana Sandri, and Sarah Maslen

3 **The Invisible Influence of Expert Consultants in Engineering Risk-Based Decision Making** 49
Jan Hayes

4 **Industrial Development, Invisible Hazards, and Bringing to Light Their Entanglements: Lessons for Ethics from the 2014 Kaohsiung Propene Explosions in Taiwan** 69
Bono Po-Jen Shih, Wen-Ling Hong, and Jr-Ping Wang

5 **Technology and Microscopy in Biomedical Education: Navigating the Ethical Landscape** 91
R. Claire Aland, Rebecca Lush, and Nicole Shepherd

6 **The Limits of Ethical Engineering Actions: The Titan Case** 115
Jan Hayes, Rune Storesund, and Sarah Maslen

7 **What Does It Take to "Settle" a Scandal? The Ethics of Hearing and Acting on Safety Concerns** 141
Dawn Goodwin and Daniel Taylor

8 **Divergent Attitudes Toward Ethics and Collective Responsibility in Canadian Engineering Discourse** 163
Kari Zacharias and Robyn Mae Paul

9 **Industry and Geopolitical Tensions: Ethical Dilemmas When Safety Meets Security** 187
Sissel Haugdal Jore, Ole Andreas Hegland Engen, Susanne Therese Hansen, and Øyvind Midttun

10 **Ethical Licensing in Practice: The Case of CRISPR Gene Editing Technology** 213
Alison McLennan and Sarah Maslen

11 **The Price of Safety: The Ethics of Occupational Risk Monetization in Brazil** 237
Carolina Bonemer Cury and Caio Caesar Dib

12 **AI as an Epistemic Technology: Navigating Ethics in Practice in Cases of Distributed Decision Making** 259
Sarah Maslen and Jan Hayes

Index 285

Notes on Contributors

R. Claire Aland is an Anatomist and Specialist Collections Officer in the School of Biomedical Sciences for a large collection of histology teaching slides at The University of Queensland, Australia. She is also the Histology Discipline Lead in the medical program, with 15 years' experience in teaching histology. Her current research interest is in the use of technology in histology teaching.

Carolina Bonemer Cury is a PhD candidate at the Faculty of Law of the University of São Paulo, Brazil. She has a Masters in Industrial and Employment Relations from the University of Turin, Italy, and a specialist degree in Labour Economics and Trade Unions from the State University of Campinas, Brazil. Her research focuses on labor law and occupational safety and health.

Caio Caesar Dib is a PhD candidate at the Faculty of Law of the University of São Paulo, Brazil, and visiting student at the Sciences Po School of Law, France. His research focuses on health comparative law, legal theory, and bioethics.

Ole Andreas Engen is Professor of Societal Safety and Security at the University of Stavanger, Norway. His research focuses on risk governance, regulation, and organizational aspects of safety and security across sectors, including energy, critical infrastructure, and climate change adaptation.

Dawn Goodwin is a social scientist and qualitative researcher working at Lancaster Medical School, UK. She teaches on aspects of social science that aid understanding of medical work as a practice, particularly around culture and safety. Her research interests focus on inquiries and investigations of harm in healthcare.

Susanne Therese Hansen holds a PhD in Political Science and is a senior researcher at the research institute NTNU Social Research, Norway. She focuses on governance, risk regulation, norms, institutions, and organizations. Her empirical research interests include critical infrastructure governance, geopolitics of energy transition, societal security, and export controls.

Jan Hayes is a Professor at RMIT University, Australia. She is a sociologist with 35 years' experience in safety and risk management. Her research interests all connect to organizational accident prevention and include professional ethics, expertise, decision making, and use of standards.

Wen-Ling Hong holds a PhD in Naval Architecture from the University of Michigan, USA. Hong is interested in engineering knowledge production and practices from a Science and Technology Studies perspective. She has studied engineering ethics, public communication of science, gender in STEM, and marine policy.

Sissel Haugdal Jore is Professor of Societal Security and Risk Management at the University of Stavanger, Norway. Her research focuses on terrorism, counterterrorism, hybrid threats, and security governance. She has served on national commissions and international expert groups and publishes in peer-reviewed journals.

Rebecca Lush is Curator of the Integrated Pathology Learning Centre and Specialist Collections Officer for histopathology slides, School of Biomedical Sciences, The University of Queensland, Australia. Her research interests include trauma-informed object-based learning and the impact of death studies on museum practice. She is currently applying such theory to expand audience groups accessing medical collections.

Sarah Maslen is a Professor at RMIT University, Australia. Her sociological research focuses on expert decision making in critical contexts, with a particular focus on the cultivation of embodied and other tacit forms of knowledge and ethics in practice.

Alison McLennan is a Senior Lecturer at the Canberra Law School, University of Canberra, Australia. She researches intellectual property and regulation of technologies and completed her PhD on the regulation of synthetic biology.

Øyvind Midttun is Communications Advisor at the Norwegian Ocean Industry Authority (Havtil), with a Master's in Risk Management and Safety Leadership and a background in journalism. He has 20 years' experience in strategic and crisis communication, and as editor of Havtil's journal *Dialogue*, he has helped shape public discourse on safety and security in the petroleum sector.

Robyn Mae Paul is Assistant Professor of Sustainable Systems Engineering at the University of Calgary, Canada. Her research and teaching focuses on applying frameworks from social justice, queer theories, and ecofeminism to broaden the narratives of engineering culture and foster more inclusive spaces and more socially just engineering designs.

Orana Sandri is a Senior Lecturer at RMIT University, Australia. Her work focuses on sustainability practices, policy, and regulation in the built environment. Through industry-based research and scholarly analysis, she examines how social, policy, and educational systems influence safety, equity, and sustainability in work practices and outcomes.

Nicole Shepherd is Lecturer in Medicine at the University of Tasmania, Australia. Her research explores the intersection of the humanities and sciences, with a current focus on narrative medicine and the ethical dimensions of healthcare. She is particularly interested in how stories shape understanding in clinical and academic contexts.

Bono Po-Jen Shih is an interdisciplinary scholar working in the areas of engineering ethics education, philosophy of engineering, and technology policy. He is currently a postdoctoral scholar at The Pennsylvania State University, USA, jointly affiliated with the Rock Ethics Institute and the Leonhard Center for the Enhancement of Engineering Education. He holds a PhD in Science and Technology Studies from The Virginia Polytechnic Institute and State University (Virginia Tech), USA.

Rune Storesund is the Director and Principal Consultant at SafeR3, a non-profit promoting state of the art research and education on safety, resilience, reliability, and risk-reduction in sociotechnical systems. Storesund holds a Bachelor of Arts in Anthropology from the University of California Santa Cruz (USA), a Bachelor of Science and Masters of

Science in Civil Engineering from the University of California Berkeley (USA), and a Doctorate from the University of California Berkeley with a focus on Civil Systems. He has over 25 years' experience studying human and organizational factors in engineered systems.

Daniel Taylor is a Clinical Research Fellow currently working at the University of Sheffield, UK. He studied medicine at Lancaster University and has research interests in systems approaches to patient safety and computational modeling of the cardiovascular system.

Jr-Ping Wang holds a PhD from the Department of Engineering Science, Oxford University, UK. He has worked on coastal hydrodynamics, flood modeling, flood risk management, and recently AI applications. Wang is interested in the role of data and simulation in engineering practices from Science and Technology Studies perspectives.

Kari Zacharias is an engineering educator and researcher with a background in Science and Technology Studies. Her work explores engineering cultures and identities, including interactions between engineering knowledges and other ways of being, knowing, and making. She is an Assistant Professor at the University of Manitoba's Department of Engineering Education, Canada.

List of Figures

Fig. 4.1	Map of explosions and casualties. (Figure courtesy of Prof. Jenq-Renn Chen)	77
Fig. 4.2	Destruction of roads and vehicles resulting from the propene explosions in Kaohsiung, 2014. (Photo by Chih Nan Fu. Courtesy of Citizen of the Earth, Taiwan)	78
Fig. 4.3	Culvert (east-west) containing the 4-inch propene pipeline (north-south). (Image courtesy of Metal Industries Research & Development Centre, Taiwan. Annotations translated by the authors)	79
Fig. 5.1	The Integrated Pathology Learning Centre, The University of Queensland, Herston campus	95
Fig. 5.2	Technology examples at the School of Biomedical Sciences, The University of Queensland, St Lucia Campus	95
Fig. 6.1	The Titan submersible. (Source: US Coast Guard Report of the Marine Board of Investigation into the Implosion of the Submersible Titan)	122
Fig. 6.2	Remotely operated vehicle image of TITAN aft tail cone. (Source: US Coast Guard Report of the Marine Board of Investigation into the Implosion of the Submersible Titan)	124
Fig. 6.3	The impact of social embeddedness and values orientation on engineering speaking up behaviors	134

List of Tables

Table 11.1 CBA clauses for collection of non-hazardous waste. Data from the collective bargaining agreement between SELUR and SIEMACO 245

Table 12.1 Reflective questions for development and use of AI technologies in decision support in critical contexts 276

CHAPTER 1

Introducing Professional Ethics in Practice

Jan Hayes and Sarah Maslen

1 The Boeing 737 MAX Case

Two Boeing 737 MAX aircraft in different parts of the world crashed within 5 months in 2018/19. Analysis of flight data soon revealed pilots in both aircraft lost the battle with aircraft control systems that insisted on lowering the nose of the aircraft, despite the pilots' attempts to correct the flight trajectory. The bizarre behavior of the technology was traced to a new safety system known as MCAS, or Maneuvering Characteristics Augmentation System, that had been installed in this new model of the well-established Boeing 737 aircraft design. The system, which pilots did not know existed, was needed to compensate for flight stability problems introduced by new, heavier engines installed in the new model. Malfunction of a single sensor in the new safety system could trigger "corrective" behavior when no problem existed, turning the aircraft into a "bucking bronco" as pilots tried to regain control (Hopkins, 2025).

From a sociological perspective, the story of the Boeing 737 MAX is often told as a tale of an organization that had lost its way (Englehardt

J. Hayes (✉) • S. Maslen
School of Property, Construction and Project Management, RMIT University, Melbourne, VIC, Australia
e-mail: jan.hayes2@rmit.edu.au; sarah.maslen@rmit.edu.au

J. Hayes, S. Maslen (eds.), *Ethics in the doing of technoscience*,
https://doi.org/10.1007/978-3-032-20002-0_1

et al., 2021; Robison, 2021). Once respected for its engineering expertise, the company had moved its headquarters from Seattle to Chicago with senior management literally distancing themselves from quality manufacturing to pursue shareholder profits. The changed company culture is often held responsible for all that followed: the decisions to re-engine the 737 for efficiency purposes, to promise airlines that no pilot recertification would be required to fly the new model, to try to overcome emerging inherent safety issues by adding complex instrumentation, to rely on a single sensor to make safety critical decisions, and to hide problems from the regulator (Majority Staff of the Committee on Transportation and Infrastructure, 2020). Digging below this organizational scale of analysis, we can appreciate how professional ethics is a key part of the 737 MAX story. In the lead up to disaster, professionals at Boeing grappled with ethical problems that they encountered within this complex social setting and made critical choices on how to navigate them.

As with any major engineering undertaking, the technoscientific work of designing and manufacturing a new aircraft involves many hands (Herkert et al., 2020). Investigations into the 737 MAX case captured stories of professionals blowing the whistle on what they observed to be ethically problematic practices due to the potential impact on passenger and crew safety. Other stories highlighted how workplace relationships primed professionals to downplay those same concerns.

Seven individuals with direct knowledge of aviation safety and Boeing's practices became whistleblowers, contacting the US government committee that was responsible for investigating the design and certification of the 737 MAX and enacting regulatory reforms designed to prevent a recurrence (U.S. Senate Committee on Commerce, 2021). One of these people, Ed Pierson, a senior plant supervisor at the 737 MAX production facility, wrote to Boeing's 737 project manager expressing his concerns that schedule and production pressures were causing quality control and safety issues. He insisted that the problems were so serious that the production line should be temporarily halted. His concerns were dismissed on multiple occasions and so he chose to retire early, 3 months before the first 737 MAX crash (Majority Staff of the Committee on Transportation and Infrastructure, 2020). The US House Committee investigation found significant value conflicts operating within the Federal Aviation Administration (FAA). Technical and safety experts identified unsafe, uncompliant practices on the part of Boeing and these problems were raised, but the Committee found numerous examples where FAA senior management sided with the airline and refused to take any enforcement action (Majority Staff of the Committee on Transportation and Infrastructure, 2020).

While Mr. Pierson and his FAA colleagues (and others who contacted the various inquiries) did their best to speak up for safety, the investigation includes many examples where professionals continued to work within a system that increasingly failed to prioritize passenger safety. One of the clearest examples relates to the actions of Authorized Representatives (AR). These individuals operate at the edge between the company and the regulator. They work within Boeing and are employed by the company but are supposed to represent the interests of the FAA and to act on the agency's behalf in confirming that Boeing systems meet regulatory requirements. The conflict of interest inherent in this arrangement meant that, in many cases, the ARs acted in Boeing's short-term interests, rather than enforcing FAA requirements. One of many examples in the 737 MAX story concerns managing data related to flight testing of the 737 MAX safety systems. Boeing data indicated that it took a test pilot more than 10 seconds to diagnose and respond to a sensor failure of the kind that caused the 2018/19 crashes. Boeing's own internal documents called this delay "catastrophic," and several ARs knew of this but did not advise the FAA. In fact, creative interpretations of certification requirements (described by one whistleblower as "slice and dice") were made with the knowledge of several ARs which allowed the MCAS system to fall outside certification requirements (U.S. Senate Committee on Commerce, 2021).

These examples, all linked to the same terrible outcomes of the loss of two passenger aircraft and everyone on board, demonstrate some of the ethical complexities and ambiguities faced by technical professionals. They must deal with conflicts between their professional sense of ethics and the values adopted by their employer. They must choose when and how to raise professional concerns and decide what to do if their concerns are not acted upon. Without the benefit of hindsight, these are difficult choices and similar decisions are made by many professionals working in technoscientific domains every day. This volume presents ten case studies of ethics in practice in the doing of technoscientific work that explore these and other questions.

To ground the coming discussions, in the following section we define key guiding parameters used in scoping this volume, including the notions of what constitutes a professional and the different ways that ethics has been conceptualized in the literature. This leads to our focus on applied ethics linked to professionals at work. Before moving to focus on professional practice, we set the scene by discussing the social nature of workplaces, decision making and ethics. We then move to three overarching themes that emerge from the following chapters. Chapters 2, 3, 4, and 5 provide a valuable commentary on how ethics is discussed (or not) in various technoscientific contexts. Chapters 6 and 7 describe how professionals navigate ethical matters in the face of organizations behaving badly, including whistleblowing. Chapters 8, 9, 10, and 11 are linked by a focus on ethics governance in various forms including regulation, professional oaths, and industry self-regulation. We conclude with some remarks about why this matters in the context of technoscientific futures, referring to Chap. 12, which addresses professional ethics in relation to artificial intelligence (AI) technologies.

2 Ethics in Professional Practice

Our particular concern in this book is professional ethics so defining what we mean by a "profession" is a reasonable place to start. In everyday conversation this term may mean simply the members of a particular occupational group. That is not the sense in which we use it. For us, a profession is "an occupational group organized to use the characteristic skill of its members for the public good" (Davis, 1988, p. 342). To be a member of a profession means to have particular technical and theoretical expertise but also to behave in accordance with agreed professional norms which are linked to the public good and may be set out in a professional code of ethics (Friedson, 2001; Middlehurst & Kennie, 1997; Sullivan, 2005). A professional is the kind of person for whom their work is more than just a job—the kind of person who would swear an oath to uphold the highest possible professional aspirations.

Many professionals work in technoscientific domains, by which we mean the multiple disciplines and settings in which there is an intertwining of scientific and technological development that shapes the world as we know and experience it (Latour, 1987). Accelerating developments in various technoscientific domains have brought with them a range of controversies. Advancements in computer science are forcing us to grapple with the multiple implications of AI technologies for how we live and work, including the changes to social relations and the nature of human decision making that such systems support. At the same time, societal attention to social justice continues to challenge long-standing professional attitudes in some sectors such as those involved in non-renewable energy production or weapons manufacturing. Professionals involved in this work grapple with micro- and macro-ethical questions, meaning those context-specific questions that emerge in their roles, as well as broader social questions as in the case of matters of social and environmental justice. Making a professionally sound choice that is in the public interest often involves negotiating competing values. Professional ethics is about how the decisions of these workers are made, the outcome chosen, and its impact. It is about what professionals ought to do (Merrill, 2011). For today's technoscientific professionals, the answers are far from clear. The pace of technology change, the extended reach of globalized work patterns, the changing global environment regarding both climate change and security, and ever-present commercial pressures challenge today's professionals in making the best choices.

Historically, ethics is the purview of philosophers with three broad schools of thought. A teleological view of ethics focuses on the consequences of action to maximize the overall benefit to self and/or others. Specifically, in a utilitarian view, ethical actions are those that provide the greatest good for the greatest number (Mill, 2011; Singer, 2011). This principle drives government spending on road upgrades to accident hotspots and spending on hospitals to communities that most need new facilities. But a utilitarian approach can be a source of tension in technoscientific work, where benefiting the public requires the actors involved to weigh the speed and direction of their work with the potential for catastrophic outcomes if it goes wrong. Uncertainty about the range of consequences and who experiences them adds further complexity as this principle is applied in fields such as gene technology, climate science, and prevention of major industrial disasters.

The primary criticism of utilitarianism is that it can lead to great suffering by a few people if this were to bring even small benefits to wider society. For example, a pure utilitarian perspective leads to challenging conclusions about the rights of animals and the severely disabled (Singer, 1993). An alternative deontological view holds that ethical considerations either permit or forbid actions based on the rightness of the action itself, not the outcome (Kant, 2017). As such, this view of ethical behavior is grounded in rules and policies. Ends do not justify means, and means can be formulated in advance as a set of prescribed ethical duties. This view is widely applied in technoscientific work with many professions and organizations developing codes of ethics setting out the ethical duties that should underpin all choices (Ersdal & Aven, 2008).

The major limitation here is that, in practice, many choices involve conflicting duties, and codes rarely address what to do in such ethically ambiguous cases. A typical example is the UK Engineering Council's Statement of Ethical Principles, which sets out four fundamental principles for ethical behavior and decision making on the part of engineers: (1) honesty and integrity; (2) respect for life, law, the environment, and public good; (3) accuracy and rigor; and (4) leadership and communication (Engineering Council, 2017). The statement acknowledges that engineering achievements can have both positive and negative impacts, but it gives no guidance on how to balance these things.

To resolve the dilemma of grounding ethics in externally imposed duties linked to either outcome or process, other ethics scholars focus on the character of individuals. In this conceptualization, judgments about character are made prior to judgments about the rightness or wrongness of actions. Theories around this form of non-rational ethics are sometimes grouped together as virtue ethics (Statman, 1997). Decisions here are driven by intuition, conscience, and faith, perhaps religious in nature but always linked to beliefs in a sense of existential virtues. In the context of technoscientific work, this perspective focuses on ethics as grounded in the values and beliefs of professionals themselves, rather than their actions. While there are contexts in which intuition, conscience, and faith are given space in professionals' ethical practice, this ethical lens underplays the degree to which ethics is largely collective in technoscientific work.

Our response to these multiple influences on professionals at work and multiple views of ethics is to study ethics in practice—from the inside and in the making (Boltanski & Thévenot, 1999; Zilberstein, 2024). Rather than taking a normative approach to either decision making or outcomes, we turn our attention specifically to professionals doing technoscientific work who, in their day-to-day practices in laboratories, clinics, offices, and data centers, navigate the progress of their respective fields. Many of them are engineers or other professionals working in established industries, some of whom must embrace major change in the face of changing societal values. Others are scientists working with new technologies that raise ethical questions that can attract significant public debate, as in the case of genetically modified food and stem cell research. This book also contains stories of medical professionals (doctors and nurses) involved in frontline roles, and educators and curators responsible for the management of biological specimens and the implications for how future generations of clinicians know and relate to the bodies of their patients.

Each chapter has at its heart an ethical dilemma. In some cases, social and professional norms conflict and, in others, judgment of what is right has changed over time. In some workplaces, ethical and moral[1] issues are discussed explicitly, but values can also remain unspoken. Taking this pragmatic approach to the study of professional ethics avoids imposing a binary of right and wrong on cases that are more helpfully seen as involving ethical ambiguity. Our approach provides lessons about what motivates technoscientific professionals and how they make ethical decisions on the ground. This provides important insights for effective professional decision making into the future as new technologies and social challenges continue to emerge, which call past ethical frameworks and practices into question.

3 The Social Nature of Professional Ethics

While the chapters in this book take a micro focus on individuals at work and their choices, it is important to acknowledge the influence of education, professional peers, and organizational context on the choices professionals make. For any individual professional, right and wrong have been learned through education and through reasoning about experiences. Ethics education in professional courses can therefore have a role in influencing engineers' moral intuition and, through that, their fluency in moral reasoning, particularly when values conflict in real-world situations (Johnson, 2020). Professional development for working professionals can have a similar role, especially where there is space for participants to reflect on the implications of course material for their professional practice (Hayes et al., 2025).

Of course, professionals' general life experience also informs their moral intuition. They do not come to their work roles as an amoral blank canvas. There may be instances in which professionals find that their moral intuitions are mutually reinforcing, though they can also be in conflict which

[1] To retain consistency with some of the ethics literature, we use "moral" only when describing the values that motivate individuals and "ethics" when describing the normative behavior of groups. As other scholars have pointed out, the ethics of individuals and the groups they belong to are far from independent and so this distinction is essentially meaningless in practice.

can lead to cases of conscientious objection, as the medical profession is currently addressing in relation to euthanasia or voluntary assisted dying legislation. Personal experiences while in professional roles can also strongly influence an individual's moral commitments. We have observed this among engineers working in hazardous industries who have previously experienced failures, which become core to their professional narratives and commitments to disaster prevention (Hayes & Maslen, 2015).

A major influencing factor is a professional's employer. The literature on organizational culture reminds us how strongly organizational norms drive the behavior of workers, with enacted (rather than espoused) ways of working becoming simply "the way we do things around here" (Schein, 1992, p. 15). In the context of high stakes decision making, organizational priorities and goals are communicated in formal processes and procedures, and also in informal practices and behavioral expectations promulgated through organizations starting from the top. As the old saying goes, "what interests my boss, fascinates me." The organizational nature of high stakes decision making has been widely used in analysis of causation of major engineering, aviation, and infrastructure disasters and also systemic failures in healthcare (Reason, 1997). Indeed, analyses of the 737 MAX crashes mentioned earlier have focused on culture change at Boeing as a key causal factor in the disasters (Hopkins, 2025). The strong focus on cost and schedule has impacted the behavior of literally thousands of employees and contractors and in many cases driven the safety-related tradeoffs they made every day. While such a viewpoint has served to extend consideration of responsibility for accidents beyond workers at the sharp end, it can be seen to cast workers as passive actors with little agency (Reason, 2008), which obscures the importance of professional ethics and its role in decision making.

Many corporate ethical decisions are related to environmental matters, particularly climate change, and so this literature provides insights into how the ethics of corporations and their employees are reconciled. Research into corporate environmentalism has highlighted the extent to which corporate decision making is increasingly linked to the impact of environmental issues on the firm itself, rather than on broader society or the planet (Du & Vieira, 2012; Wright & Nyberg, 2017; Wright et al., 2021). Corporate self-interest regarding economic growth and profitability has led to actions targeting external legitimacy and influence in the eyes of policy makers and the public (Du & Vieira, 2012). Studies of how corporate environmentalism impacts employees are much rarer despite the fact that employees find themselves at the very heart of a paradox whereby companies that owe their success to carbon-intensive activities must now transition away from them (Demers & Gond, 2019; Halttunen et al., 2022; Littel et al., 2025). All such strategies have an element of neutralizing employees' moral responsibility, but most "want to see themselves as a force for good and believe in their ability to create change within the organization" (Halttunen et al., 2022, p. 8). Thus, employees sometimes stay despite value conflicts with their employer.

Another important form of ethical socialization is professional networks which often address ethics both formally and informally. Sometimes ethical requirements of a profession are not aligned with those of employers. For instance, Bolton (2000) captured how gynecology nurses offered gestures of care beyond their job description, but in keeping with their professional codes of conduct. For many professions, ethical expectations are communicated via a professional code of practice or similar. Even if not used as an oath, many professions have published codes of ethics, and in some cases, compliance is legally enforced via professional licensing and registration (e.g., for medical doctors and airline pilots). Consistent with a deontological view of ethics, written frameworks are an important influence on professionals and are often the subject of academic study (see Regulation and Governance section below). But to be a professional is also to be part of a social group and the informal workplace interactions that go with that. Empirical work with practicing professionals has shown the extent to which behaviors are passed on informally to more junior members of the profession (Iedema et al., 2009; Maslen & Hayes, 2020).

The cases examined in the following chapters illustrate the interplay between these various motivating factors and how professionals navigate the ethical demands made of them. All our chapters are about professionals acting within a social setting driven by the norms of their professions, organizations, industries, legal systems, societies, and nations. Our focus is not to examine this social context alone but to dig into the strategies of ethical engagement that professionals adopt in the course of their work (Maslen et al., 2021).

4 Talking about Ethics

Across many technoscientific contexts, there has been a common position that scientific work is driven by facts and figures and so is value-free. In this way of thinking about technical professions, all technical decisions are objective and unbiased. Conflicting priorities are often resolved by consideration of risk, which in itself is seen as an objective exercise (Aven, 2025; Kastenhofer, 2009). Such a view excludes ethics which, by its very nature, is grounded in judgment and so subjective. Despite this intransigent positivist framing of technical work, the role of expertise, experience, and judgment in professional decision making is well established (Goodwin, 1997; Groopman, 2000; Hayes et al., 2021, 2025; Trevelyan, 2014). Technical competence is critical for professional practice, but it is not enough to ensure that professionals drive technology-based choices in a socially acceptable way. Ethics is about how technical competence is applied in specific contexts taking into account the long-term consequences of the range of technological possibilities (Johnson, 2020).

A lack of engagement with ethics in relation to technology means that ethics conversations within technoscientific organizations, at least in the English-speaking world, are often minimized (Molthan-Hill, 2014). Describing this phenomenon, Bird and Waters (1989) adopted the term "moral muteness," i.e., a reluctance for technical and management professionals to describe their decisions in moral terms, even when there are clear moral dimensions. Baden and Harwood (2013) described how the ethical concept of corporate social responsibility has been co-opted by economic considerations and language. Similarly, Cech (2014) observed a "culture of disengagement" within engineering degree programs that serves to destabilize, rather than clarify, the ideological commitments of students. To make matters complicated, a lack of moral language does not necessarily mean that ethics is ignored. Our previous work on engineers who make high-stakes decisions on matters of public safety found that the engineers use technical, action-oriented language to discuss decisions, but their concerns are grounded in an ethics of public safety. (Maslen et al., 2021)

Other research focuses on moral muteness as present in communications between businesses and clients. We see this particularly clearly in relation to tourism ethics. Vespestad et al. (2025) showed that online travel agents conceal unsustainable tourism practices to fragile destinations by using pro-environmental and euphemistic language. Conversely, Font et al. (2017) describe "greenhushing" practices where sustainable tourism operators make public only around one third of their sustainability-related actions. In both cases, businesses are seeking to gain financial advantage from potential customers by avoiding conflict between their values and those they perceive might be held externally to the organization.

Several chapters in this volume demonstrate how the broader sector orientation toward ethics impacts how it is dealt with. The analysis of the Grenfell Tower apartment building fire in which 72 residents died (Chap. 2) shows how important it is for ethical conversations to be part of professional practice, particularly in work environments that are geographically and temporally distant from the consequences of choices made. In this case, two companies manufactured flammable building products and sold them as suitable for use in refurbishment of high-rise residential buildings, with communications between these companies and their clients being morally mute regarding the safety of building residents. Despite regulations that should have prevented the use of flammable products in this service, many individuals over many years chose to prioritize profit over the safety of residents. Hayes, Sandri, and Maslen relate this to widespread moral disengagement within these organizations that reframed wrongdoing as acceptable or even the right thing to do. Countering such a strong cultural orientation requires individual ethical practice that identifies moral disengagement antecedents, rejects moral muteness, and exercises rigorous due diligence that keeps safety paramount in the face of commercial pressures.

In Chap. 3, Hayes describes a different working environment where long-term safety risks are more explicitly discussed, but conversations remain largely technical. In the world of hazardous infrastructure, risk is never zero so a key question is "how safe is safe enough?" Hayes shows that for the Australian pipeline sector, despite formalized processes in place, the answer to this question is ultimately a matter of professional judgment in the face of uncertainty. Further, specialist consultants employed by the companies who own and operate the infrastructure drive expenditure recommendations based on their values. This is underacknowledged in the industry where a positivist orientation drives the prevailing framing that the consultants are applying a deterministic process to reach an answer.

Acknowledging the cultural context to engineering ethics means we must also recognize the ways that ethics may manifest differently around the world. In Chap. 4, Shih, Hong, and Wang take us into the development of engineering ethics in Taiwan. Through their examination of the 2014 Kaohsiung propene explosions, they demonstrate the extent to which the condition of the pipeline infrastructure was ignored by the responsible engineers despite the potential impact on public safety. While professional ethics is seen significantly as an individual concern in the

West, Shih and colleagues describe a collective ethics at work in Taiwan. They postulate that the collective focus on the benefits of industrial development under martial law (pre-1987) left a long tail in engineering ethics. Until the 2014 accident, public awareness of the danger of pipelines was very low. Despite an increased focus on engineering ethics in professional education, the lack of action by those responsible for pipeline integrity reflects the societal priorities established pre-1987 in which the danger from such infrastructure was hidden from view. They see the "demand" factor of public expectations as the key determinant of professional ethical behaviors.

In contrast to these engineering cases where ethics is rarely discussed, in the working environment described by Aland, Lush, and Shepherd in Chap. 5, the ethical principles informing decision making are a regular topic of conversation. They take us into the world of museums and collections that contain human specimens collected over many decades. As educators they find an ethical need to reconsider how specimens are used. They are faced with a need to balance the uncertain provenance of some of the collection with changed societal expectations regarding consent and modern educational imperatives for learning materials reflecting the diversity of society. Digitalization of the collection further complicates the issue by giving users greater agency. To address these challenges, they have developed and adopted a new professional ethical framework in their work. The history of the collection is acknowledged rather than hidden, and users (collections professionals, educators, researchers, and students) are encouraged to draw on operationalized principles of respect, dignity, and reverence as new items are added to the collection and older ones are gradually withdrawn and retired.

Taken together, these cases demonstrate the importance of explicit discussion of ethical issues to ensure that professional practice is driven not only by short-term, expedient needs of a given employer but includes mindful consideration of the long-term and broader goals of a given profession in the wider interests of society. This requires dealing with uncertainty, being open to change (e.g., as new data becomes available), and being able to balance conflicting values. The cases show that context plays an important part in determining the best way forward. However, avoiding ethical conversations (as in the Grenfell and Kaohsiung cases) or delegating ethics questions to others (as in the case of the Australian pipeline sector consultants) is not a robust way forward.

These are not only historical issues or questions relevant to established technologies. The collections case emphasizes that processes such as digitalization extend the reach of individual ethical choices, making explicit discussion prior to action even more essential. Against this extended reach of decisions, the development and deployment of AI technologies in decision support requires that we engage with the ethical implications of their use, which are far reaching. AI technologies are positioned as "objective" within the public imagination in comparison to the judgments of human actors (Burrell & Fourcade, 2021), which has the potential to further obscure conversations about professional ethics in technoscience just at the point they are becoming most critical.

5 Organizations Behaving Badly

As we have described earlier, the ethical norms of professionals are at least partly a reflection of their employing organization, but there are also situations in which individual and organizational ethical norms collide. When individuals find themselves in a serious value conflict with their employer, there are three options available to them. Some people choose to live with the cognitive dissonance that results from such a conflict. Others choose to leave the organization quietly and find employment elsewhere. A third group chooses to blow the whistle on what they see as unacceptable behavior on the part of their employing organization. In the context of corporate wrongdoing, we are perhaps most used to seeing organizations and even governments called out for unethical behavior in relation to human rights and financial crimes such as fraud, but whistleblowing is also relevant to technoscientific professionals who observe, for instance, a divergence from what they understand to be competent and thus ethical practice.

The term "whistleblowing" stems from the historical practice of police officers blowing a whistle when they saw a crime being committed to draw attention to the wrongdoing (Johnson, 2020). The act of calling out organizational actions in such a strident way is a major departure from the collaborative, learning orientation of normal professional behavior. The adversarial nature of whistleblowing is reflected in the way whistleblowers are often treated. Medicine has a well-established whistleblower tradition to address patient safety concerns by reporting co-workers to authorities. Despite this, the term is seen as negative, and blowing the whistle is seen as risky behavior (Jones & Kelly, 2014). In their study of internal whistleblowing in the Norwegian oil and gas industry, Stene and Eikemo (2019) found that colleagues of the whistleblower were sometimes unsupportive for fear of sanctions. The most commonly reported retributive actions taken against whistleblowers were "the undermining of the whistleblower as a person in public, assignment of fewer or less attractive tasks, bullying and threats, creation of a disciplinary case, or denial that notification has even occurred" (Stene & Eikemo, 2019, p. 3907). It is little wonder, therefore, that many professionals choose to move on rather than face such opposition.

Two chapters deal with the response of professionals when their assessment of what is right deviates so far from the actions of those around them that they choose to formally report their concerns outside the organization—with very different results. In Chap. 6, on the Titan submersible implosion, Hayes, Storesund, and Maslen explore the limits of engineering responsibility in a corporate setting. In 2023, the OceanGate submersible Titan imploded during a deep ocean dive to the wreck of the Titanic, killing the five crew, including the OceanGate CEO. The CEO had insisted over several years that the design of the vessel's hull was safe, despite significant evidence presented by technical personnel to the contrary. Using this case, the authors explore the actions of professional engineers in raising safety concerns. Because of their actions in speaking up, several key professionals were pushed into leaving the organization. The analysis shows that this move from organizational insider to outsider is driven by their orientation toward professional ethics but also by their degree of social embeddedness in the organization. This has implications for professionals in considering whether and when they are willing to blow the whistle to try to raise the alarm about rogue organizations and CEOs.

This interplay between those with safety concerns and a broader organizational system that does not want to listen is also the topic of Goodwin and Taylor's Chap. 7. They recount the case of overuse of diamorphine in an elderly care unit (Gosport War Memorial Hospital) between 1987 and 2001, which has been estimated to have shortened the lives of 450 patients. Using theory on epistemic injustice, they analyze how this practice was allowed to continue for such an extended period when both nurses and patients' families had raised issues, firstly with the hospital itself and then with the police. Their concerns were dismissed, partly because of the epistemic privilege enjoyed by medical practitioners who see those who raise safety concerns as troublemaking whistleblowers, rather than professionals (in the case of nurses) trying to improve outcomes. Goodwin and Taylor conclude that for things to change in the healthcare sector, medical and management professionals must embrace an ethical responsibility to value wider sources of knowledge than is currently often the case.

These cases from very different sectors show some common themes. Initially, professionals raised their concerns internally with a focus on performance improvement, but rather than being thanked for their initiatives, in both cases, the professionals who stood up for safety were marginalized and/or excluded in various ways. In neither case did whistleblowers effect immediate change. In both the Titan and Gosport cases, people died when they should not have. It took several years for complaints to be taken seriously, but in both cases significant systemic improvements were identified because of their efforts. These cases thus raise important questions regarding how professional concerns are treated, both within organizations and by external governance mechanisms. It is common in the aftermath of disaster to find that concerns were raised, only to be ignored. The increasing rate of technological change and the broadening reach of high-stakes decisions mean that organizational learning is as critical as ever and all sources of knowledge are a gift that should not be dismissed.

6 Regulation and Governance

In professions that hold a more positivist view of professional knowledge, there is often a view that professional ethics can be completely codified and so ethics governance is a matter of compliance with written rules, codes of ethics, and any laws that may apply. Even in our broader framing of ethics as a social practice which is both determined and constrained by norms at organizational, professional, and societal levels, codes of ethics have an important role. They have several advantages. First, they can bring issues into the open and allow debate and agreement. They can even serve as a "prop" in social interactions to navigate conflicts (Maslen & Ransan-Cooper, 2017). They also make it clear to newcomers what is expected and remind established members of a profession what is required of them. No one is going to argue with the position that designing faulty equipment through ignorance, taking bribes, or misrepresenting research results is wrong. But modern practice often requires professionals to balance conflicting values (Maslen, 2019). Ethical frameworks can facilitate discussion about ethics in ways that can resolve conflict, achieve compromise, and allow ethical dilemmas to be acknowledged.

Despite these major advantages, codes of ethics have limits too as the case of the Hippocratic Oath shows. The Hippocratic Oath is the most well-known professional code of ethics as sworn by medical practitioners in many parts of the world as part of their entry into the profession. Originating in ancient Greece, the oath is often summarized as an exhortation for medical practitioners to do no harm, although this expression has much more recent origins (Smith, 2005). Modern medical scholars contend that the Hippocratic ideal of medical practice focused on a holistic model of healthcare and applied ethical rules that are still valid today (Kleisiaris et al., 2014).

Despite the overall support for the relevance of the oath within the profession, commentators describe the difficulties in deciding what it means for medical ethics in modern times (Askitopoulou & Vgontzas, 2018a). There are debates around the contemporary meaning of specific clauses such as whether the prohibition of "destructive pessaries" in the oath is best interpreted as anti-abortion or rather anti-gynecological harm (Menéndez Motta, 2024). Another clause prohibits medical practitioners from giving deadly drugs which some have used as an argument against involvement from qualified doctors in capital punishment, voluntary assisted dying, or abortion. One passage sets out a medical practitioner's duty to fully share medical information with other clinicians. This has been said to be an obligation that can be brought forward from ancient times with little difficulty (Miles, 2005). But even this belies restrictions raised by copyright and other issues linked to the modern value of intellectual property. Other scholars argue that the primary message of the oath is the primacy of the doctor-patient relationship and the essential focus on the patient's best interests, rather than prohibition of any specific form of treatment (Askitopoulou & Vgontzas, 2018b).

As Miles (2005) notes, values change over time, but the Hippocratic Oath remains beneficial for the profession in its role in opening ethical debates in contemporary medical discourse. The changing nature of the healthcare sector and an increasing focus on profits has also led some doctors to reflect whether recitation of the oath is a form of professional hypocrisy in an age of medicine as a business (Bakshi, 2019). Ethics in practice is still the subject of contemporary research in medicine, as in the recent study by Archibong and Ukeagu (2025) on the moral, emotional, and psychological implications for medical practitioners of the ethical conflict between patient confidentiality and holding knowledge that is potentially life-changing for other people.

Despite the long-standing use of the Hippocratic Oath, medicine has not been without ethical controversies. Having a written code of ethics does not preclude the medical profession from instances of behavior deemed to be unethical, even by contemporaneous standards. The well-known Tuskegee Syphilis Study is one example where African American men were exposed to syphilis, resulting in many additional deaths among the patients and their immediate families. The study ran from 1931 to 1972 and was eventually halted only when a whistleblower leaked the story to the press (Gardner & McClelland, 2025).

A code of ethics in the form of an oath can only ever expound general principles and as we see there is plenty of scope for disagreement when it comes to grounding meaning in specific daily practices. Nevertheless, support for the Hippocratic Oath remains strong among medical professionals (Askitopoulou & Vgontzas, 2018a, 2018b; Sioutis et al., 2021). Despite its weaknesses, the strengths of a professional oath are such that there have been calls from other professions to follow the medical example (Abdulla, 2001).

Another key point of contention regarding the content of professional codes of ethics is their scope. Ethics scholars are broadly in agreement that values driving individual behaviors such as honesty and fair dealing are legitimate ethical considerations for a professional code. There is a growing debate in engineering ethics in particular as to whether individual ethics should extend beyond such micro-ethical concerns to focus on macro-ethical questions (McAninch, 2023). The fundamental issue is whether individual professionals have an ethical responsibility to address the big issues facing society that involve social and/or policy matters relevant to an entire profession (e.g., response to climate change or how to deal with AI). Engineering ethics scholars are increasingly arguing that this is the case. As Skoglund and Böhm (2019) have noted, the separation between inside and outside of organizations is becoming harder to maintain in an era of increasing employee activism and we see that throughout this book.

Chapter 8 by Zacharias and Paul on Canadian engineers' views on revisions to the iron ring ceremony (an optional ritual welcoming new engineers to the profession which includes reciting an ethical oath) speaks directly to the question of the engineering profession's readiness to address the macro-ethical issues inherent in the technological challenges of the future and public expectations. Historically, the ceremony was grounded in an oath to act honestly and to maintain the reputation of the profession (among other things). Online debates about the merits of updating the oath to include a commitment to considering the broader consequences of their professional activities have strong support from some engineers while others see this as an unwelcome politicization of their profession. Engineers of this view see the profession as currently apolitical with a cultural responsibility to uphold a set of micro-ethical professional traditions such as the original version of the oath. Others see the importance of language in the oath itself and creating a text that is more inclusive of racial and gender diversity in the engineering profession. Another group sees that the profession has a collective responsibility to consider different perspectives, which encourages them to identify and resolve ethical dilemmas posed by their work in a more mindful way.

Another theoretical touchpoint here is the literature on regulatory compliance. Regulatory compliance is often a strong professional value grounded in the belief that regulation represents societal expectations of what constitutes ethical behavior. The reality of corporate life also means that the requirements of various regulatory regimes are often conflicting, and yet most of the literature focuses on compliance with a single regulatory goal (Chester & Hayes, 2023). Haines and Gurney's (2003) study of the ideological conflict between consumer protection legislation and workplace health and safety is a rare study of regulatory conflict in practice. Similar to Chester and Hayes's (2023) study on the conflict between technical safety and economic regulation in the gas utility sector, they found that conflict was glossed over by regulators and resolution was delegated to companies and ultimately technical staff who were responsible for meeting both sets of requirements in practice.

Chapter 9 from Jore and colleagues addresses a specific case of conflicting ethical imperatives driven by changing governance arrangements initiated at a national level. The Norwegian offshore oil and gas sector has a long-standing reputation for having an effective safety culture reflected by a generally low accident rate. A significant factor in this is the tripartite model uniting government, employers, and the unions in acting for a common goal of a safe workplace. Given the changed national security situation, petroleum companies have now become part of the system ensuring national security. Jore and colleagues argue that this requires different ethical behaviors that no longer link to long-established values such as trust, transparency, inclusion, and fairness. The challenge for both workers and organizational systems is to navigate these tensions without eroding foundational principles which support the democratic and moral fabric of the sector.

In Chap. 10, McLennan and Maslen do not deal with conflicting regulatory goals but rather show what happens when a new technology remains insufficiently regulated. They describe the ethical implications of the current dispersed governance arrangements for gene editing technology, specifically CRISPR-Cas 9. This technology is widely used in health, animal sciences, and agriculture with varying ethical implications. The sector operates under an arrangement known as "ethical licensing" where the ethics of potential uses of the technology become one factor in the patent holder's decision whether or not to license any given applicant. But McLennan and Maslen show that in practice this system is vague in terms of the ethical principles involved, encourages scientists to find work-arounds rather than engaging with the ethical principles, and imposes restrictions at the commercialization stage of new applications, rather than when the technologies are being considered. Especially in the face of new gene editing techniques driving new patents, both scientists and society need more legitimate governance approaches grounded in open discussion and sound ethical principles.

The approach to risk monetization used for regulation of worker safety in hazardous industries in Brazil demonstrates another way in which governance can fail when it comes to ethics. In Chap. 11, Cury and Dib describe the daily ethical dilemmas created when low-paid workers working in dangerous conditions are paid a premium in the form of hazard pay, i.e., increased income to compensate for increased workplace risk. In this situation, what does it mean for union personnel to represent the interests of workers? What is the role of workplace health and safety staff? This market-based approach was arguably intended to give employers a financial incentive to make workplaces safer, but in practice this merging of financial motivations with the goal of sending workers home safely to their families is an ethical disaster.

Together, these cases demonstrate that governance arrangements can drive professional behaviors, but not always as intended. Whether we are reading the literature on codes of ethics or regulatory compliance, there is a gap between the aspirations set out in governance documentation and the messy reality of professional life. Our case studies demonstrate that ethics as a matter of general principles is often a far cry from ethics in practice where conflicting values pull professionals in different directions. The case of workplace safety legislation in Brazil shows that regulations intended to drive behaviors can have unintended consequences at odds with the initial intent. This does not mean that the best option is to leave ethical decisions to individual professionals. The gene editing case shows that if governance arrangements are lacking, professionals will find a way around them to act in accordance with their own assessment of what is right. The Norwegian case provides further guidance on good ethics governance, by showing that governance arrangements need to be flexible as priorities change. Finally, we see that professionals value governance arrangements to provide guidance, but they must have legitimacy within the profession as the case of the Canadian iron ring oath demonstrates.

7 Why Does this Matter?

Beneath the banality of everyday work life, technoscientific professionals are making judgments that impact the lives of others. Be it the safety of engineered systems or the speed and accuracy of medical diagnoses, the wellbeing of animals subject to genetic modification or the fairness and respect paid to historical specimens of human origin, mindful, ethical choices contribute to the society we want to live in. Despite the fundamental importance of these decisions, we see that they are being made sometimes without explicit discussion of the values involved. In many ways the most surprising thing about our case studies taken as a whole is the minimal explicit consideration of ethics in practice across workplaces in so many different parts of technoscience. The gap between formal codes of professional ethics and the reality of everyday work is filled by each individual—sometimes successfully but sometimes not. In general, the case studies demonstrate that organizations tend to avoid explicit discussion of the values on which their systems and employee expectations are founded. Further, governance processes, such as they exist, tend to articulate general principles removed from the messiness of everyday work which leaves professionals on their own, particularly when values come into conflict.

These issues are important when it comes to the best uses of existing technologies for the overall benefit of society, but they come into even sharper focus when we consider the escalating use of AI. The idea of AI as machines capable of reasoning is both captivating and terrifying. Chapter 12 concludes this volume by addressing the ethical significance of the widespread use of AI as an epistemic technology. The very reach of the technology means it has the potential for huge benefits in efficiency and accuracy but also the potential for significant harm when things go wrong. Chapter 12 addresses ways in which AI changes the epistemology that surrounds decision making practices by exacerbating problems of distance between decision makers and impacts, raising new questions about how to learn about rare events, and creating conditions for professionals to rely too heavily on the technology. Mechanisms for raising concerns are also currently lacking. Surrounding all these considerations is the question of responsibility for outcomes when humans and machines are working together. This chapter calls for a refocusing on the epistemic practices of professionals, as a counter to what we see as too strong of a focus on the affordances of the technology.

More broadly, drawing on all the cases presented here we see the value of observing ethics in the decision making moment, not in the principles (operationalized or otherwise) that dominate ethics governance today. We call for more empirical studies of ethics in practice across the range of technoscientific professions in order to provide a richer set of cases. The individual studies we present here illustrate some methodological approaches. Some take as their object of study a specific disaster case. Rather than undertaking a broad study of causation, these chapters focus in on the ethics of the choices made by key individuals in context. Such cases provide an important opportunity for learning what happens when ethical decision making fails, the circumstances that lead to such failures, and how they might be avoided. In the tradition of normal operations studies in safety, other chapters use interviews supported by analysis of written artifacts to dig into the ethics of everyday decision making. Netnography offers yet another approach to appreciate the ethical debates that professionals have with one another outside of a research setting. These qualitative approaches show that rich values-based data are available if only researchers pay attention to them. We emphasize that the value of such cases lies in their nuanced exploration of the navigation of ethical questions within technical disciplines, especially when considered through a social science lens. They hold significance for both academics seeking a greater understanding of this field and governance bodies grappling with how to make principles-based frameworks relevant. Most importantly, cases support technoscientific professionals seeking to improve their ethical practice.

References

Abdulla, S. (2001). An oath for scientists? *Nature*. https://doi.org/10.1038/conference010222-18

Archibong, E. I., & Ukeagu, L. O. C. (2025). The Hippocratic oath, principle of confidentiality and the moral burden of physicians. *IBOM Medical Journal, 18*(3), 502–515. https://doi.org/10.61386/imj.v18i3.709

Askitopoulou, H., & Vgontzas, A. N. (2018a). The relevance of the Hippocratic oath to the ethical and moral values of contemporary medicine. Part I: The Hippocratic oath from antiquity to modern times. *European Spinal Journal, 27*(7), 1481–1490. https://doi.org/10.1007/s00586-017-5348-4

Askitopoulou, H., & Vgontzas, A. N. (2018b). The relevance of the Hippocratic oath to the ethical and moral values of contemporary medicine. Part II: Interpretation of the Hippocratic oath—Today's perspective. *European Spine Journal, 27*(7), 1491–1500. https://doi.org/10.1007/s00586-018-5615-z

Aven, T. (2025). Risk science and politics: What is and should be the relationship? *Risk Analysis, 45*(2), 274–282. https://doi.org/10.1111/risa.16558

Baden, D., & Harwood, I. A. (2013). Terminology matters: A critical exploration of corporate social responsibility terms. *Journal of Business Ethics, 116*(3), 615–627. https://doi.org/10.1007/s10551-012-1498-9

Bakshi, A. (2019). *Hippocratic oath or hypocrisy? Doctors at crossroads*. SAGE Publications India Pvt, Ltd.

Bird, F. B., & Waters, J. A. (1989). The moral muteness of managers. *California Management Review, 32*(1), 73–88. https://doi.org/10.2307/41166735

Boltanski, L., & Thévenot, L. (1999). Sociology of critical capacity. *European Journal of Social Theory, 2*(3), 359–377. https://doi.org/10.1177/136843199002003010

Bolton, S. C. (2000). Who cares? Offering emotion work as a 'gift' in the nursing labour process. *Journal of Advanced Nursing, 32*(3), 580–586. https://doi.org/10.1046/j.1365-2648.2000.01516.x

Burrell, J., & Fourcade, M. (2021). The society of algorithms. *Annual Review of Sociology, 47*, 213–237. https://doi.org/10.1146/annurev-soc-090820-020800

Cech, E. A. (2014). Culture of disengagement in engineering education? *Science, Technology, & Human Values, 39*(1), 42–72. https://doi.org/10.1177/0162243913504305

Chester, L., & Hayes, J. (2023). Regulatory conflict and a latent public safety risk? The case of gas infrastructure. *Law & Policy, 46*(1), 63–86. https://doi.org/10.1111/lapo.12231

Davis, M. (1988). Professionalism means putting your profession first. *Georgetown Journal of Legal Ethics, 2*, 341–357.

Demers, C., & Gond, J.-P. (2019). The moral microfoundations of institutional complexity: Sustainability implementation as compromise-making at an oil sands company. *Organization Studies, 41*(4), 563–586. https://doi.org/10.1177/0170840619867721

Du, S., & Vieira, E. T. (2012). Striving for legitimacy through corporate social responsibility: Insights from oil companies. *Journal of Business Ethics, 110*, 413–427. https://doi.org/10.1007/s10551-012-1490-4

Englehardt, E., Werhane, P. H., & Newton, L. H. (2021). Leadership, engineering and ethical clashes at Boeing. *Science and Engineering Ethics, 27*(12), 1–17. https://doi.org/10.1007/s11948-021-00285-x

Engineering Council (2017). Statement of Ethical Principles for the Engineering Profession. Engineering Council and Royal Academy of Engineering. London

Ersdal, G., & Aven, T. (2008). Risk informed decision-making and its ethical basis. *Reliability Engineering and System Safety, 93*, 197–205. https://doi.org/10.1016/j.ress.2006.12.018

Font, X., Elgammal, I., & Lamond, I. (2017). Greenhushing: The deliberate under communicating of sustainability practices by tourism businesses. *Journal of Sustainable Tourism, 25*(7), 1007–1023. https://doi.org/10.1080/09669582.2016.1158829

Friedson, E. (2001). *Professionalism: The third logic*. University of Chicago Press.

Gardner, U. G., & McClelland, S. (2025). Peter Buxtun, the Tuskegee study whistleblower: Memorandum of a hero. *Journal of the National Medical Association, 117*(1), 32–33. https://doi.org/10.1016/j.jnma.2025.01.004

Goodwin, C. (1997). The blackness of black: Color categories as situated practice. In L. B. Resnick, R. Säljö, C. Pontecorvo, & B. Burge (Eds.), *Discourse, tools and reasoning: Essays on situated cognition* (pp. 111–140). Springer.

Groopman, J. (2000). *Second opinions: Stories of intuition and choice in the changing world of medicine*. Viking Penguin.

Haines, F., & Gurney, D. (2003). The shadows of the law: Contemporary approaches to regulation and the problem of regulatory conflict. *Law & Policy, 25*(4), 353–380. https://doi.org/10.1111/j.0265-8240.2003.00154.x

Halttunen, K., Slade, R., & Staffell, I. (2022). "We don't want to be the bad guys": Oil industry's sensemaking of the sustainability transition paradox. *Energy Research & Social Science, 92*, 102800. https://doi.org/10.1016/j.erss.2022.102800

Hayes, J., & Maslen, S. (2015). Knowing stories that matter: Learning for effective safety decision-making. *Journal of Risk Research, 18*(6), 714–726. https://doi.org/10.1080/13669877.2014.910690

Hayes, J., Maslen, S., Holdsworth, S., & Sandri, O. (2021). Defining the capable engineer: Non-technical skills that support safe decisions in uncertain, dynamic situations. *Safety Science, 141*, 105324. https://doi.org/10.1016/j.ssci.2021.105324

Hayes, J., Maslen, S., Holdsworth, S., & Sandri, O. (2025). The part that we play: Engineers' perceptions of their responsibility for public safety. *Safety Science, 187*, 106862. https://doi.org/10.1016/j.ssci.2025.106862

Herkert, J., Borenstein, J., & Miller, K. (2020). The Boeing 737 MAX: Lessons for engineering ethics. *Science and Engineering Ethics, 26*, 2957–2974. https://doi.org/10.1007/s11948-020-00252-y

Hopkins, A. (2025). *Boeing, the 737 MAX crisis and aviation safety*. CRC Press.

Iedema, R., Jorm, C., & Lum, M. (2009). Affect is central to patient safety: The horror stories of young anaesthetists. *Social Science & Medicine, 69*(12), 1750–1756. https://doi.org/10.1016/j.socscimed.2009.09.043

Johnson, D. G. (2020). *Engineering ethics: Contemporary and enduring debates*. Yale University Press.

Jones, A., & Kelly, D. (2014). Whistle-blowing and workplace culture in older peoples' care: Qualitative insights from the healthcare and social care workforce. *Sociology of Health & Illness, 36*(7), 986–1002. https://doi.org/10.1111/1467-9566.12137

Kant, I. (2017). *Groundwork for the metaphysics of morals: With an updated translation, introduction, and notes*. Yale University Press.

Kastenhofer, K. (2009). Debating the risks and ethics of emerging technosciences. *Innovation: The European Journal of Social Science Research, 22*(1), 77–103. https://doi.org/10.1080/13511610902770594

Kleisiaris, C. F., Sfakianakis, C., & Papathanasiou, I. V. (2014). Health care practices in ancient Greece: The Hippocratic ideal. *Journal of Medical Ethics and History of Medicine, 7*, 1–5.

Latour, B. (1987). *Science in action: How to follow scientists and engineers through society*. Harvard University Press.

Littel, F., Yu, A., & Rodgers, P. (2025). Enacting aspirational modes of being: Oil and gas employees' subject formation and telos under corporate environmentalism. *Organization Studies, 46*(10), 1353–1377. https://doi.org/10.1177/01708406251357860

Majority Staff of the Committee on Transportation and Infrastructure. (2020). *Final committee report: The design, development and certification of the Boeing 737 MAX*. https://transportation.house.gov/imo/media/doc/2020.09.15%20FINAL%20737%20MAX%20Report%20for%20Public%20Release.pdf

Maslen, S. (2019). Safety management through values: A critical engagement with the moral labor of disaster prevention. *Safety Science, 120*, 484–491. https://doi.org/10.1016/j.ssci.2019.07.045

Maslen, S., & Hayes, J. (2020). "This is how we debate": Engineers' use of stories to reason through disaster causation. *Qualitative Sociology, 43*, 191–212. https://doi.org/10.1007/s11133-020-09452-1

Maslen, S., Hayes, J., Holdsworth, S., & Sandri, O. (2021). When ethics is a technical matter: Engineers' strategic appeal to ethical considerations in advocating for system integrity. *Science and Engineering Ethics, 27*(4), 46. https://doi.org/10.1007/s11948-021-00324-7

Maslen, S., & Ransan-Cooper, H. (2017). Safety framing and compliance in relation to standards: Experience from the Australian gas pipeline industry. *Safety Science, 94*, 52–60. https://doi.org/10.1016/j.ssci.2016.12.011

McAninch, A. (2023). Go big or go home? A new case for integrating micro-ethics and macro-ethics in engineering ethics education. *Science and Engineering Ethics, 29*(3), 20. https://doi.org/10.1007/s11948-023-00441-5

Menéndez Motta, M. (2024). *The commentary on the Hippocratic oath by Johann Heinrich Meibom: Medicine and philology in the early 17th century*. FAU University Press. https://doi.org/10.25593/978-3-96147-706-7

Merrill, J. C. (2011). Theoretical foundations for media ethics. In A. D. Gordon, J. M. Kittross, J. C. Merrill, W. Babcock, & M. Dorsher (Eds.), *Controversies in media ethics*. Taylor & Francis Group. https://doi.org/10.4324/9780203829912-2

Middlehurst, R., & Kennie, T. (1997). Leading professionals: Towards new concepts of professionalism. In J. Broadbent, M. Dietrich, & J. Roberts (Eds.), *The end of the professions? The restructuring of professional work* (pp. 50–68). Routledge.

Miles, S. H. (2005). *The Hippocratic oath and the ethics of medicine*. Oxford University Press.

Mill, J. S. (2011). *Utilitarianism*. Andrews UK.

Molthan-Hill, P. (2014). The moral muteness of managers: An Anglo-American phenomenon? German and British managers and their moral reasoning about environmental sustainability in business. *International Journal of Cross Cultural Management, 14*(3), 289–305. https://doi.org/10.1177/1470595814532076

Reason, J. (1997). *Managing the risks of organizational accidents*. Ashgate.

Reason, J. (2008). *The human contribution: Unsafe acts, accidents and heroic recoveries*. Ashgate.

Robison, P. (2021). *Flying blind: The 737 MAX tragedy and the fall of Boeing*. Penguin Random House.

Schein, E. (1992). *Organizational culture and leadership* (2nd ed.). Jossey-Bass.

Singer, P. (1993). *How are we to live: Ethics in an age of self-interest*. Random House Australia.

Singer, P. (2011). *Practical ethics*. Cambridge University Press. https://doi.org/10.1017/CBO9780511975950

Sioutis, S., Reppas, L., Bekos, A., Limneos, P., Saranteas, T., & Mavrogenis, A. F. (2021). The Hippocratic oath: Analysis and contemporary meaning. *Orthopedics, 44*(5), 264–272. https://doi.org/10.3928/01477447-20210819-08

Skoglund, A., & Böhm, S. (2019). Prefigurative partaking: Employees' environmental activism in an energy utility. *Organization Studies, 41*(9), 1257–1283. https://doi.org/10.1177/0170840619847716

Smith, C. M. (2005). Origin and uses of primum non nocere—Above all, do no harm! *The Journal of Clinical Pharmacology, 45*, 371–377. https://doi.org/10.1177/0091270004273680

Statman, D. (1997). *Virtue ethics: A critical reader*. Edinburgh University Press.

Stene, L. K., & Eikemo, I. (2019). Is whistleblowing a safety mechanism or a threat? A study of petroleum-related companies in Norway. In *Proceedings of the 29th European Safety and Reliability Conference (ESREL 2019)*. Research Publishing Services. https://doi.org/10.3850/978-981-11-2724-3_0578-cd

Sullivan, W. M. (2005). *Work and integrity: The crisis and promise of professionalism in America*. Jossey-Bass.

Trevelyan, J. (2014). *The making of an expert engineer*. CRC Press.

U.S. Senate Committee on Commerce, Science, and Transportation (2021) Aviation Safety Whistleblower Report, United States Senate, Washington

Vespestad, M. K., Hehir, C., & Koivunen, K. (2025). How moral disengagement links to destination marketing organisations' moral muteness in their sustainability communications. *Journal of Sustainable Tourism, 33*(1), 9–27. https://doi.org/10.1080/09669582.2023.2276034

Wright, C., & Nyberg, D. (2017). An inconvenient truth: How organizations translate climate change into business as usual. *Academy of Management Journal, 60*(5), 1633–1661. https://doi.org/10.5465/amj.2015.0718

Wright, C., Nyberg, D., & Bowden, V. (2021). Beyond the discourse of denial: The reproduction of fossil fuel hegemony in Australia. *Energy Research & Social Science, 77*, 102094. https://doi.org/10.1016/j.erss.2021.102094

Zilberstein, S. (2024). Ethical dilemmas and collaborative resolutions in machine learning research for health care. *Socius: Sociological Research for a Dynamic World, 10*, 1–16. https://doi.org/10.1177/23780231241259671

CHAPTER 2

"Keep as 'VERY CONFIDENTIAL!!!!'": How the Behavior of Technical Professionals Contributed to the Grenfell Tower Disaster

Jan Hayes, Orana Sandri, and Sarah Maslen

1 Introduction

On June 14, 2017, an electrical fault in a fridge/freezer on the fourth floor of the high-rise residential Grenfell Tower in central London started a small fire in an apartment kitchen. The fire spread very quickly to engulf the entire building and 72 people were killed. The rapid and unexpected escalation of a minor domestic fire to a national catastrophe occurred because of a recent refurbishment of the tower block that had clad the building in insulation and rain screen cladding, both of which were highly flammable. Rainscreen cladding systems have become popular in the United States and Europe in the last two decades, driven by increasing requirements for building envelopes that enhance thermal performance and thus reduce carbon footprint. These systems are more energy-efficient

J. Hayes (✉) • O. Sandri • S. Maslen
School of Property, Construction and Project Management, RMIT University, Melbourne, VIC, Australia
e-mail: Jan.hayes2@rmit.edu.au; orana.sandri@rmit.edu.au; sarah.maslen@rmit.edu.au

J. Hayes, S. Maslen (eds.), *Ethics in the doing of technoscience*,
https://doi.org/10.1007/978-3-032-20002-0_2

than the external skin of buildings built in the twentieth century (Grauer, 2019) and also update the look of a building to current tastes. The Grenfell Tower was subject to refurbishment of this kind. Building materials, especially those used in high-rise residential buildings, are supposed to be fire resistant and the materials used in the Grenfell Tower refurbishment had fire certificates that seemed to indicate that they met UK building standards. At the time, UK standards for fire rating of such materials were less stringent than the comparable European standards, but the materials used on Grenfell failed to meet even these lower standards.

This case thus raises significant ethical questions regarding the behavior of the many actors involved in the development, testing, and certification of the material, as well as that of the architects, construction contractors, building inspectors, and fire risk assessors involved in the retrofitting of the material to Grenfell Tower. This complex network of construction professionals provided numerous opportunities for critical questions to be asked about the fire rating of the materials, particularly given the growing warnings posed by building cladding fires in the years before 2017. However, the fraudulent nature of the claims did not come to light until after the fire. To tell the whole story in one book chapter would not do justice to the nuances. We thus focus on the people involved at the beginning. As we will note in our conclusion, there were unethical behaviors on the part of the people involved in the selection of the cladding for Grenfell who simply did not do their due diligence, but in our view the most brazenly negligent behavior came from the people dealing with the testing of the external cladding (produced by Arconic) and the insulation (produced by Celotex) as part of the fire certification process. Our detailed examination of the case thus focusses on the actions of individuals within the two companies that manufactured the cladding and the insulation. We seek to understand what led these technical professionals to behave as they did over an extended period and what lessons this has for other technical professionals about ethical decision making.

Our analysis shows how fire certificates were brought into existence, allowing these dangerous materials to be promoted for use across the UK. Our analysis is based on the work of the public Grenfell Tower Inquiry which investigated the detailed course of events of the night of the fire (Phase 1) and how it was possible that such an event could happen, despite what were thought to be effective regulations (Phase 2). We draw on the Inquiry reports (Grenfell Tower Inquiry, 2024b, 2024c), transcripts of evidence from witnesses, and key documents all available in the

Inquiry archive.[1] When it comes to details of the cladding installed on the Tower, the Inquiry sought oral evidence from six witnesses employed by Arconic. After extended legal arguments, only two people (UK residents) gave oral evidence. The other four potential witnesses (European residents) chose not to give oral evidence and the Inquiry had no power to compel them. Some of these people provided written witness statements. In contrast, five key people from Celotex involved in the testing and subsequent marking of the insulation panels used on Grenfell Tower gave oral evidence. We show how these organizational settings institutionalized deception on critical safety matters through moral disengagement and loss of safety imagination. Our analysis shows that this is not a case of "bad apples", but rather such attitudes were widespread within the construction industry. This case has important lessons for all technical professionals in seeing the potential for such attitudes in other contexts. Identifying moral disengagement antecedents, rejecting moral muteness, and exercising rigorous due diligence that keeps safety paramount in the face of commercial pressures are key behaviors that individual professionals can adopt to challenge such attitudes before they become entrenched.

2 Administrative Evil and Moral Disengagement

Each piece of major engineered infrastructure such as a dam, a transportation system, or a high-rise building is a complex assemblage of thousands of components. Each component is produced by many organizations and individual professionals. The responsibility for getting things right is thus collective, as key safety decisions are made based on knowledge communicated across a complex network of organizations and individuals, typically over long temporal scales. How then can individual professionals act ethically within this social system of which they are a part?

The concept of the "golden thread" (Hackitt, 2018) captures the need for key safety requirements to be specified at an early stage of development and for such requirements to be mindfully enacted through all parts of the system so that the end product, in this case a refurbished high-rise building, meets all safety requirements during its ongoing life as a home for hundreds of residents. While maintaining the golden thread is critical to ensure safe outcomes, research into engineering responsibility has shown

[1] The archive of the inquiry website is available here: https://webarchive.nationalarchives.gov.uk/ukgwa/20250319155650/, https://www.grenfelltowerinquiry.org.uk/.

how difficult this can be. With "many hands" (van de Poel et al., 2012) involved in the work, tasks can become compartmentalized and it is possible that no-one feels a forward-looking responsibility for the cumulative consequences of work (van de Poel, 2011). In other words, individuals can lack a "safety imagination" (Pidgeon & O'Leary, 2000), i.e., the ability to link their day-to-day actions to safety outcomes. This is exacerbated by the distance, both temporal and geographic, that may exist between actions and the ultimate outcomes (Hayes et al., 2021).

One theorization of how technical professionals can contribute to "evil" acts during their seemingly ordinary work is the notion of "administrative evil" first articulated by Adams. Starting with the Holocaust and extending to other more recent policy failures, the concept of administrative evil in public administration explains why "ordinary people, within their normal professional and administrative roles, can engage in acts of evil without being aware that they are doing anything wrong" (Adams, 2011, p. 275). When the values that drive action become inverted, people may even view their evil activity as good. Recently, researchers have lamented that while the concept "offers explanatory power to understand and describe why and how people [civil servants] behave badly and even unethically in organizational contexts" (Virtanen et al., 2022, p. 1), few scholars have applied it to broader organizational situations. A few rare exceptions have used administrative evil to study finance sector failures (Dillard & Ruchala, 2005; Lilly et al., 2021), but the concept remains unused in the context of safety-related disasters. While its application is rare, we believe the notion has some resonance in our case, and so we position our analysis in this context.

Administrative evil occurs "when the routine administrative tasks carried out by employees result in great harm to others" (Lilly et al., 2021, p. 588), but this does not mean that the individuals themselves are evil. As Adams (2011, p. 277) articulates:

> Ordinary individuals are simply acting appropriately in their organizational roles — in essence, just doing what those around them would agree they should be doing — and at the same time, participating in what a critical and reasonable observer, usually well after the fact, would call evil.

Scholars advancing these theorizations have also identified organizational conditions that foster the development of organizational evil. In Adams's (2011) view, technical rationality epitomized by a scientific-analytic mindset combined with a belief in technological progress can become powerful enablers of administrative evil but, as Landry (2009) says, professional expertise is also a factor. In these paradigms, the tools and techniques of professions are applied in the service of an agreed goal which, in the case of technoscientific work, may be public safety. While such an approach is often successful, when the goal is unclear, technical rationality flounders (Landry, 2009). As described above, there may be significant separation between actions and safety outcomes and, in such cases, organizational and professional cultures can resort to appeals to instrumental goals such as profitability or technical progress, while ignoring longer-term or more distant consequences. Lilly et al. (2021, p. 593) cite an example of this from a Wells Fargo employee who told the Inquiry into the fake accounts scandal:

> I am NOT writing this letter to bring an investigation on my store, my district, my region-that is not where the root of the problem lies. It lies on upper management who has increased the goals to the "must cheat to achieve" level.

Employees can assume that extreme behaviors are acceptable or even encouraged when employers impose rigid measures of success that seem unachievable by any other means (see also Hayes et al., 2024).

Actions linked to administrative evil often involve lying on the part of individual employees, particularly for defensive purposes, i.e., to avoid conflict and mediate power relationships, or for material reasons, i.e., concealing damaging information for personal or corporate gain (Jenkins & Delbridge, 2020). Following Barnes (1994), a lie is an action intended to deceive others. This differs from sharing untruths unknowingly; the awareness of and intent to deceive is key. Additionally, a lie may involve deliberately not acting (for instance, failing to pass on information) also with the purpose of deceiving. Despite some exceptions, in Western culture, lying is seen as wrong because honesty is viewed as a moral virtue (Hayes et al., 2024). As a result, for administrative evil to become entrenched in workplace behaviors, other mechanisms are needed to deal with the cognitive load imposed by the conflicting values that result, particularly when evidence of the adverse consequences starts to accumulate, i.e., workers want to see themselves and their choices in a positive light despite the link to bad outcomes.

Moral disengagement theory describes eight psycho-social mechanisms where people can "commit transgressive acts without experiencing personal distress" (Bandura et al., 2000, p. 58). These are:

- Moral justification (give behavior a moral purpose).
- Euphemistic labelling (to mask ethical problems).
- Advantageous comparison (everyone does it or worse).
- Displacement of responsibility (others made me do it).
- Diffusion of responsibility (it was all of us).
- Disregarding or distorting the consequences (it's not so bad really).
- Dehumanization (of those impacted).
- Attribution of blame (we were provoked).

If these mechanisms become entrenched in the attitudes of organizational actors, the severity and extent of lying can increase beyond the control of any one individual or group. As this occurs, "more organizational members will be persuaded, enticed, coaxed, threatened or socialized to join in the deception" (Fleming & Zyglidopoulos, 2008, pp. 841–842). The deception itself can act to bind those involved together (Shilling & Mellor, 2015) and those who do not wish to engage in the deceit can choose to leave (Grover, 2005). Language also plays an important role here. Other ethics researchers have noted that the language in corporations rarely engages with moral issues directly (Bird & Waters, 1989; Jordan, 2009; Maslen et al., 2021), potentially contributing to moral disengagement.

Bandura et al. (2000) discuss how mechanisms of moral disengagement were evident in prominent cases of corporate ethical wrongdoing including Bhopal, Ford Pinto, Nestle, and Three Mile Island. While there are clear organizational responsibilities to prevent such toxic cultural characteristics from developing, individual technical professionals are not completely without agency. Little previous research has engaged with this directly so our aim here is to go beyond questions of whether technical professionals are bad apples and/or passive actors to focus on what it means for professionals to behave ethically when faced with such organizational cultural characteristics.

3 Arconic's Cladding Fire Certificate

The rainscreen panels used in the new external wall of Grenfell Tower were a product called Reynobond 55 PE manufactured in France by Arconic Architectural Products (Arconic). These panels are of aluminum composite material (ACM) construction, consisting of two thin sheets of aluminum with a polyethylene (PE) core. Arconic sold this product in flat sheets which were then customized during installation according to the designer's specifications for a given application. Certificates detailing the fire safety performance of the product were required to sell the product for use in cladding high-rise buildings. For this reason, the results of fire safety testing were of keen interest to Arconic senior management and technical and sales personnel.

3.1 Early Fire Testing

Early fire testing of the Reynobond 55 PE panels raised issues in both the testing process and the interpretation of results by Arconic. Reynobond 55 PE panels were first fire tested in 2004 by the French agency Centre Scientifique et Technique du Bâtiment (CSTB) with sample specimens in two typical fabricated arrangements, one cut for riveting and one folded into cassettes. The riveted specimens passed testing for use in high-rise residential buildings (Test 5A). However, there were later indications of test manipulation based on an internal email from Arconic's Head of Technical Sales Support Team, Mr. Wehrle, who described the result as "'arranging' for the system to pass" (Grenfell Tower Inquiry, 2024e, p. 16). This also suggested that the results were not truly representative of how the riveted panel system would behave in practice. The first test of the cassette system (Test 5B) had different results and was aborted due to its flammability. Records show that no further tests were carried out. Despite the failed outcome of the 5B Test, as Mr. Wehrle stated in evidence, Arconic professionals involved in testing chose to believe that the cassette test was a rogue result and that they were entitled to claim that the results for the riveted panel test were representative of all uses of Reynobond 55 PE. Arconic subsequently used these results based on their interpretations to market Reynobond 55 PE panels in Europe for high-rise buildings.

3.2 A UK Fire Safety Certificate Is Produced

While Arconic was using marketing material in Europe based on the one successful CSTB test of the riveted panels (Test 5A), they were preparing for a major push into the UK market and realized that they would need a local fire rating certificate to convince the UK market to use their products. They chose the British Board of Agrément (BBA) as their certifying agency. Arconic's initial application to BBA for certification, prepared by Mr. Wehrle, included the results of the CSTB Test 5A and some other tests done in the UK on different cladding products. The failed CSTB Test 5B was not included in the application despite BBA's standard contract for certification which requires applicants to provide results of all relevant fire tests over the previous three years. Test 5B fell within this timeframe and the Inquiry found that Mr. Wehrle's omission of this test from Arconic's application "can only have been deliberate" (Grenfell Tower Inquiry, 2024e, p. 22).

Overall, the Inquiry found that Mr. Wehrle "deliberately and dishonestly concealed from the BBA the fact that … [Reynobond 55 PE's] reaction to fire was significantly affected by the fixing system adopted" (Grenfell Tower Inquiry, 2024e, p. 21). In late 2007, the fire safety certificate for Reynobond 55 PE was drafted by the BBA based on data provided by Arconic and without requiring any additional fire tests. Mr. Wehrle and Arconic's UK sales representative Mr. Southgate reviewed drafts of the certificate on at least four occasions between October 2007 and January 2008. Multiple wording changes were suggested by them and adopted by BBA. The final certificate issued in January 2008 expressly states that the product meets UK regulatory requirements when used in either riveted or cassette form.

3.3 Hiding the Problem

From 2007 onward, Arconic was actively promoting their products on the UK market with the BBA certificate playing a key role. Despite the claim that their product met all safety requirements, some employees (including Mr. Wehrle) were becoming aware of flammability issues with PE cladding based on several building fires, the changing regulatory environment (where ACM with a PE core was being banned in some European countries), and the actions of their competitors in this changing market.

The Inquiry found an internal email exchange in 2010 where Mr. Wehrle admitted to an Arconic sales executive Ms. Moyses and Arconic Sales Manager Mr. Scheidecker that the cassette form of Reynobond 55 PE does not meet fire regulations and "we have to keep as 'VERY CONFIDENTIAL!!!!'". Mr. Scheidecker responded: "This shouldn't even have been mentioned" (Grenfell Tower Inquiry, 2024e, p. 36). Also in 2010, a customer insisted on seeing certification for the fire performance of the system in cassette form. After internal discussion in which Mr. Wehrle noted they were "not clean" on this matter, Mr. Wehrle sent the customer the riveted test results, implying that the cassette fire performance would be better. Mr. Wehrle did not give oral evidence to the Inquiry, but a colleague who did acknowledged that Mr. Wehrle had lied to a client about the fire performance of the cassette form.

Two high-rise building fires in Dubai within six months in 2012/13 highlighted the problem with ACM panels with a PE core. Anticipating questions from concerned clients about the flammability of Arconic products, a UK sales representative, Ms. French initiated an email chain that included Arconic's president, Mr. Schmidt. A discussion ensued about whether Arconic should move to sell only fire-resistant panels, but no action resulted. Salespeople continued to promote the use of the same kind of panels that had caused these two major fires.

3.4 *More Testing and Reclassification*

In 2013, the CSTB was again commissioned to test Reynobond 55 PE panels in both riveted and cassette configurations. These tests were initiated by Mr. Wehrle to resolve long-standing issues with the test status of cassette-installed panels. Again, the cassette test had to be abandoned, but this time the riveted panel test also produced unsatisfactory results. In the end, both arrangements were given a fire rating by CSTB that meant the panels could not meet UK fire regulations in any configuration. In early 2014, Mr. Wehrle issued an instruction to the sales team, telling them that all Reynobond 55 PE products had been reclassified and could no longer be sold for installation on high-rise residential buildings in the UK or Europe.

BBA again had an opportunity to correct the fire compliance certificate when it came up for renewal in 2013/14. However, Arconic did not respond to BBA's repeated requests for results of any additional testing since the 2006 tests on which the 2008 certificate was based. The BBA

Project Manager for the 2013/14 certificate renewal, Maria Barbeito, was a new graduate materials engineer with no experience in cladding or building regulations. She and her supervisor continued to press Arconic for information, receiving no response. In January 2015, after numerous requests for data had been ignored for over 16 months, senior management at BBA decided to reissue the fire certificate for Reynobond 55 PE based on 2006 data. The Inquiry described this action as "inexplicable" (Grenfell Tower Inquiry, 2024e, p. 56). The BBA had the option to suspend the certificate, but this was apparently never considered. Overall, the Inquiry noted that BBA's weak processes and attitudes led what many in industry assumed to be a robust governance body to become "the victim of dishonest behavior on the part of unscrupulous manufacturers" (Grenfell Tower Inquiry, 2024d, p. 15).

With the UK fire certificate remaining in place, Ms. French continued to use it to promote the use of Reynobond 55 PE panels to clients in the UK despite the contradictory instruction from the technical team via Mr. Wehrle. It was around this time that the contract for use of Reynobond 55 PE panels in cassette form was secured for the Grenfell Tower refurbishment project. The Inquiry report pointed out that it was not an oversight; rather, "it reflected a deliberate strategy to continue selling Reynobond 55 PE in the UK based on a statement about its fire performance that [Arconic] knew to be false" (Grenfell Tower Inquiry, 2024d, p. 13).

4 Celotex

FR5000 (later rebranded as RS5000) is a polyisocyanurate (PIR) thermosetting foam thermal insulation designed for use in rainscreen cladding systems to provide a thermal layer between the cladding and the wall. Manufactured by Celotex from 2008, it was widely used as part of the rainscreen cladding system on the Grenfell Tower refurbishment. Like Arconic's Reynobond 55 PE, Celotex's RS5000 marketability and use on buildings above 18 m was dependent on certification that it was a "material of limited combustibility" and conformed to building regulation fire safety guidance material that allowed the product to be used on high-rise buildings such as Grenfell Tower, or that it was part of a cladding system that satisfied fire performance criteria BR 135 for external thermal insulation for walls in multi-story buildings. Achieving such certification of RS5000 was central to Celotex's strategy to enter the high-rise cladding market.

4.1 *"Above 18 m Project": Prioritizing Marketability and Profit*

As rainscreen cladding systems grew in popularity for high-rise buildings, Celotex sought to enter the market. In 2011, they began tracking sales inquiries for insulation for buildings over 18 m and recognized that they were losing sales, particularly to their competitor Kingspan (Grenfell Tower Inquiry, 2024a), because they lacked a suitable product for these applications. Head of marketing, Paul Evans, investigated testing requirements in late 2012 and tasked Product Manager, Jonathan Roper, with developing, testing and certifying an above 18 m insulation product. Mr. Roper, a recent business graduate with no relevant prior experience or qualifications, was closely managed by Mr. Evans and assisted by Jamie Hayes who was technical support personnel but also lacked relevant expertise and qualifications. The inquiry later suggested that Mr. Roper was likely assigned the task because of his youth and inexperience.

Around this same time, Saint-Gobain Construction Products UK Ltd. acquired Celotex and set an expectation for a 15% increase in profits. With this budgetary driver and evidence of 70,000 m^2 loss of insulation sales due to the lack of certification, Mr. Roper at the request of Mr. Evans developed the "Above 18 m Action Plan" (Grenfell Tower Inquiry, 2024a) in March 2013.

4.2 *Mirroring Kingspan's Misleading Approach*

To begin the Above 18 m project, Mr. Roper began researching products and testing systems to certify FR5000 for use at height. Mr. Roper's research concluded that Kingspan had taken advantage of a single, unrepresentative product test and carefully worded BBA certificate and marketing materials to promote their K15 insulation panels for use in buildings above 18 m. In a report to Mr. Evans, Mr. Roper highlighted a poor market understanding of fire regulations and fire performance which Kingspan exploited. Mr. Roper believed that only building control officers might challenge K15's use, a concern Kingspan effectively neutralized through specific wording in the K15 certification documents. In a report to Mr. Evans in November 2013, Mr. Roper then concluded: "Trying to do the right thing requires a complete re-education of the market and this would require a huge campaign and probably a lawsuit" (Grenfell Tower Inquiry, 2024e, p. 128). The Inquiry indicated that Mr. Roper believed the proper approach was to sell the product only for use in the tested system and to inform specifiers about the improper use of Kingspan K15 to avoid further market distortion. In response to Mr. Roper's report, Mr. Evans replied, "We are trying not to create a 'me too' here but if we do, it will be for the right reasons" (Grenfell Tower Inquiry, 2024e, p. 128). The Inquiry found that Mr. Evans most likely meant that if they decided to call Kingspan out for their behavior, then it would be to create a commercial advantage for Celotex. In the end, Celotex management decided to mirror Kingspan's testing and marketing approach. Mr. Roper and Mr. Evans then met with fire testing consultants and the certifier, Building Research Establishment (BRE), to determine the best testing system to ensure FR5000 would meet the relevant performance criteria.

4.3 *Manipulated Testing*

BRE's first test of FR5000 undertaken in February 2014 failed and was stopped after 26 minutes when flames overtopped the rig. Mr. Clark, the Burn Hall Manager at BRE who observed the test, said that he had "seen worse fails" (Grenfell Tower Inquiry, 2024e, p. 131) and made suggestions to Celotex to improve the system. Despite the significant failure and associated costs of testing, Celotex pursued a second test due to the substantial commercial value of certification. According to the Inquiry, "the discussions centered on how to ensure that any further test could be designed to be successful" (Grenfell Tower Inquiry, 2024e, p. 132). The test rig was redesigned by Mr. Roper, Mr. Evans, and Mr. Hayes, with what the Inquiry described as a conscious attempt to manipulate the test so the system would pass. This included thickening the rainscreen panels from 8 to 12 mm, the inclusion of a magnesium oxide layer placed at the top of the rig to help fire resistance and prevent flames from overtopping the rig and then concealing this added layer with a thinner, 8 mm cladding. Ventilation gaps were also reduced in the second rig. No new drawings of the second system were produced until after the test.

The second rig passed a test in May, 2014, meeting the BR 135 performance criteria. Mr. Clarke from BRE, who also observed the second test, wrote up a draft report detailing the second rig design, test conditions, and outcomes. While the report did include a photo of the test rig being dismantled that showed the magnesium oxide layer, the descriptions within the report made no mention of this added layer. NHBC, the largest private building control body in the country, also reviewed the draft test report and raised concerns that the rig had been deliberately over-engineered using thicker cladding panels with minimal gaps. In response, Mr. Roper falsely claimed that only the thicker cladding panels had been available for the testing rig. Following directions from Mr. Evans and other colleagues, Mr. Roper then requested that BRE remove the photo showing the magnesium oxide layer from the report and supplied updated drawings that added a ventilation gap not in the tested rig and that also omitted the magnesium oxide layer.

In August 2014, BRE issued the classification report based on the test report. The certification was 12 pages and contained the same test terms as the test report; however, it omitted photographs, the inclusion of the magnesium oxide layer and use of the thinner, 8 mm cladding. Based on this certification, RS5000 also received certification from Local Authority Building Control (LABC) in the UK, who assess certain products for compliance with aspects of the Building Regulations. This certification was used as further evidence to market the RS5000 for use in cladding systems over 18 m.

4.4 Misleading Marketing and Exploiting Market Confusion

With the certification report now available, Mr. Roper and Mr. Evans created an abridged version for their customers, omitting details of the test rig—a similar approach taken by Kingspan. This abridged version had no diagrams or photographs of the system tested, only a list of components that omitted the 6 mm magnesium oxide boards and the 8 mm ruby Eternit boards. This was sent to clients in August 2014. Later, a customer who complained about the lack of detail in the abridged version was sent the certification report and was told it had the full details.

With certification complete, Mr. Roper then developed a marketing plan that presented RS5000 as Celotex's "primary rainscreen application offering to compete directly with Kingspan K15 & Rockwool Duo-Slab" (Grenfell Tower Inquiry, 2024e, p. 145). The report also claimed that it was "the first PIR insulation board tested and approved to BR 135 and therefore acceptable for use in buildings above 18m in height" (Grenfell Tower Inquiry, 2024e, p. 145). These phrases were used repeatedly in future marketing materials for RS5000.

Part of marketing RS5000 was targeting companies that lacked technical knowledge of fire rating and certification processes and implications. The Inquiry found that a recently employed sales manager Mr. Roome had shared materials with colleagues within Celotex categorizing potential customers into three tiers: Tier 1 were aware of problematic testing and interrogated the test results of materials they used; Tier 2 included customers who were open to considering the use of combustible insulation; and Tier 3 included customers who opted for combustible insulation, partly because they were unaware of the 18 m restriction. This document indicates that Celotex understood the varying levels of technical knowledge among potential buyers and recognized that the market for RS5000 would likely be restricted to those who lacked expertise or expert advice, or who did not comprehend or investigate the fire performance of RS5000. It was this marketing strategy, targeting customers with weaker technical understanding, that led to the sale of RS5000 for installation on Grenfell Tower.

5 Discussion

5.1 Profit/Market Share Driving Individual Actions

The Inquiry findings (Grenfell Tower Inquiry, 2024a, 2024e, 2024f) show that key staff involved in the testing and marketing of both Reynobond 55PE cladding and RS5000 insulation were acting with the economic interests of their employers (Arconic and Celotex, respectively) in mind. Acting in the interests of one's employer to increase profitability is a legitimate ethical goal in the sense that this is what many professionals are paid to do with a view to providing shareholder returns and a vehicle for ongoing employment among other legitimate corporate goals. What makes these cases exceptional is the extent to which this goal overtook any other ethical considerations. Other typical professional values include compliance with the law, honesty, and consideration of the public good (Middlehurst & Kennie, 1997), but these values were sidelined within these two companies, and as soon as those constraints on professional behavior were gone, conditions for administrative evil existed. Legitimate behaviors to maximize profitability then included misrepresenting fire test results, lying to customers and lying to certification agencies. We see that lies told in the service of profitability then took two main forms. Some lies were falsehoods, directly materially motivated by corporate gain. In many cases these were combined with defensive lies where key details were concealed in order to avoid conflict (Jenkins & Delbridge, 2017).

For example, Arconic's decision to hide unfavorable fire test results of the cassette form of Reynobond 55 PE was so extensive that in giving evidence, Arconic's president Mr. Schmidt "accepted that if the fire had not taken place it was likely that [the cassette test results from 2004] would have remained secret" (Grenfell Tower Inquiry, 2024e, p. 17). Similarly, all those involved with FR/RS5000 at Celotex intended to deceive others about the fire rating of the panels through blatant misreporting of the design of the test rig that passed and also failing to pass on information about the real test conditions in marketing and information materials. The actions undertaken by employees at Arconic and Celotex clearly show how their routine administrative tasks led to great harm through subsequent loss of life in the Grenfell Tower fire (Lilly et al., 2021). Such actions could be seen as "mindless", as a result of deceptive practices, later becoming habits and thus taken for granted (Ashforth & Anand, 2003, p. 15).

5.2 *Organizational Context*

While those involved in certification and marketing lied for commercial gain, Inquiry evidence shows that this was not just a one-off or the work of a small group of rogue employees in each company but rather an organizational way of working. The singular focus on profitability was driven by management and grounded in commercial reasons. Arconic was keen to break into the UK market and saw the need for a fire safety certificate for their product that implied it met UK regulations as a key part of that campaign. Similarly, Celotex launched the "Above 18 m Project" because it was losing sales to the rival company Kingspan, particularly in the high-rise insulation market, and needed a product for use above 18 m. Given these strong organizational motivations linked to commercial survival and growth, employees at both companies manipulated fire tests and concealed unfavorable data to obtain certification, ensuring their products could be marketed as safe for high-rise use despite failing fire tests. The safety of those who might live or work in buildings that had their products attached was not an organizational consideration.

Lying at Arconic and Celotex became a collective, shared practice beyond the control of any one individual. At Celotex, new employees like Mr. Roper were essentially "socialized to join in the deception" (Fleming & Zyglidopoulos, 2008, p. 842). Mr. Roper quickly learned from senior staff to conceal test rig modifications (e.g., hiding magnesium oxide boards) and mislead BRE and customers. At Arconic, employees were warned not to speak even internally about the failed fire test of the cassette system. Significant lies do not arise randomly within organizations; instead, they stem from deeply rooted cultural traditions and social standards associated with partial truths and the act of "playing along" (Hodgson & Balmer, 2022). At Celotex, management reinforced these practices, creating an organizational environment where deception was a deliberate strategy and so from its inception lying was accepted by employees as an intrinsic work feature central to the organization's commercial success (Jenkins & Delbridge, 2017). At Celotex, there was even explicit consideration as to whether they should act in compliance with regulations and call out their competitor (Kingspan) for the misleading approach they took to testing and marketing their own insulation product. Instead of challenging Kingspan's deceptive claims, Mr. Roper concluded it was easier to copy them than to "re-educate the market".

5.3 *Justifying Evil*

Celotex's attitude to legal compliance shows the process of moral disengagement at work (Bandura et al., 2000). Some Celotex staff justified the manipulation of fire tests by pointing to industry practice—essentially claiming that because Kingspan was doing it, they had to do it too. Later, at the Inquiry, they sought to shift responsibility for flammable insulation being installed on the Grenfell Tower onto others in the supply chain for not checking RS5000's suitability. A similar pattern can be seen at Arconic where the serious fire problem with the cassette form of Reynobond 55 PE was hidden from certifying agencies and customers over more than a decade.

These practices fit within three of Bandura's eight "psycho-social mechanisms":

- Euphemistic labelling, as in the case where Arconic's Mr. Wehrle recast the fire test stopped due to the danger posed by the highly flammable product as "not a test" because, according to the relevant standard, a completed test requires three samples. This justified not sharing the results.
- Advantageous comparison, which is evident in Mr. Roper's "no alternative" response and reliance on Kingspan's misleading fire tests and marketing. This justified Celotex's lies as a forced response to

low industry standards. Staff had the choice to report concerns to regulators but instead chose commercial advantage.

- Lastly, displacement of responsibility, which is evident at the Inquiry, where staff from both manufacturing companies blamed others in the supply chain (architects, contractors, installers) for not checking material suitability. This was also evident at Arconic when Mr. Wehrle said he would have given the BBA the results of Test 5B if they had asked for them, although how they were supposed to know to ask for the results of a test that they did not know had been performed was not addressed.

Taken together, these interconnected behaviors reveal a progression. Lies began as materially motivated acts for profit. They evolved into collective organizational practices, strategically deployed as business tactics that were normalized and reinforced by mis-selling within regulatory gaps.

It is perhaps important to note that there is no evidence of explicit discussion of the flammability of the cladding and insulation material as having ethical consequences in either company. Indeed, we could describe these companies as "morally mute" (Maslen et al., 2021). In both companies, decisions were consistently framed as "commercial" or "technical", rather than ethical. Marketing strategies that targeted uninformed customers were presented as legitimate sales tactics, rather than unethical exploitation. The language of "technical compliance" thus supported the normalization of deception by concealing its moral implications.

5.4 *Just a Few Bad Apples?*

Looking from the outside, it is tempting to see either the individuals involved or the manufacturing companies themselves as bad apples. Comparing their actions to what would be required from professionals who see the broader moral context of their actions leads us to the conclusion that these groups and individuals were lacking. Yet for this disaster to take place, other lines of defense held by other organizational and professional actors also needed to fail. The Arconic fire compliance certificate renewal points to how the negligence of other organizations and individual professionals allowed the continued use of the cladding material, as we saw in the "inexplicable" reissuing of the certificate by BBA after Arconic did not provide the testing data necessary for certification. A factor here is inexperience on the part of the personnel involved but also presumably an urge to close out a job to the degree that due diligence receded from importance.

As we foregrounded in the introduction, the actors directly involved in the Grenfell Tower project are also not innocent. We came into this analysis assuming that those professionals who worked on Grenfell had been taken in by falsified fire ratings. What we found instead were professionals who did not even engage with them. There were many stages in the design, installation, and inspection processes surrounding the project where the dangerous nature of these materials could have been brought to light, despite the fraudulent fire certificates in circulation for both the cladding and the insulation. As architect, Studio E was responsible for the design of the external wall of the building and for the choice of the materials used in its refurbishment. They originally planned to use non-flammable zinc cladding and Celotex insulation, knowing it was widely used in this type of application and so assuming it was compliant with fire regulations. In 2013, under pressure from the building owners, Studio E was looking for options to reduce the cost of the work. One option that came to light was changing from zinc exterior cladding to ACM. They struck up a relationship with Harley, a cladding supplier and installer, who promoted use of Reynobond ACM panels as the cheapest option. Reading all the detailed correspondence over six months, the Inquiry noted that the key decision making factors were a tradeoff between aesthetics and cost with no discussion of fire safety. Studio E produced the final project specification without investigating the fire performance of the proposed materials. They apparently assumed that the products recommended by the vendors were appropriate based on their use in similar applications elsewhere. The selection of insulation material for use on the refurbished tower followed a similar pattern.

6 Conclusion

Administrative evil is overwhelmingly an organizational phenomenon and as such it is common to assume that corporate misbehavior starts at the top, but researchers of business ethics have shown that the interactions between leaders and followers regarding ethics are more complex. Blaming the boss or formally displacing responsibility is a mechanism that "allows individuals to allocate outside themselves (upwards to the leader), the responsibility for whatever morally questionable actions they may take"

(Almeida et al., 2021, p. 549). Thus, taking an exclusively organizational lens plays into one of the psycho-social mechanisms of moral disengagement: others made me do it.

While we do not question the often overwhelming strength of one's organizational context on the ethical scripts available to technical professionals (Maslen et al., 2021), we nonetheless consider it to be important to reflect on the degree to which individuals can influence their organizational circumstances and act to discourage unethical behaviors from growing up and becoming established. The Grenfell Tower case suggests three potential strategies. The first is to develop an awareness of the potential for moral disengagement among professionals. Whenever the justification for strategic decisions becomes one or more of Bandura et al.'s (2000) eight mechanisms, professionals should take a step back and consider the broader ethical implications of the proposed actions, and in particular the potential for longer term consequences that breach values such as honesty, obeying the law, or public good.

Individual professionals can also reflect on the consequences of their behaviors and choices at work and encourage their colleagues to do the same. In the Grenfell Tower case, actors across the organization were chasing a business outcome without genuine consideration of the real-world consequences of sending a flammable product to market or installing a flammable product on a building. In short, they lacked a safety imagination (Pidgeon & O'Leary, 2000). This does not mean continually insisting that the sky is falling but rather avoiding moral muteness (Bird & Waters, 1989; Maslen et al., 2021). We call on professionals to establish and protect a clear ethical line in the sand.

Ethical action is also a matter of due diligence. There were many occasions between the development of the building materials and their installation in which the actors involved were effectively rubber stamping those communications that came across their desk without deeply engaging in the material itself. Banal and routine tasks such as opening email attachments and checking technical details become core professional activities once professionals are mindful of the consequences of errors and their forward-looking responsibility (van de Poel, 2011) to do their part to prevent catastrophic outcomes.

References

Adams, G. B. (2011). The problem of administrative evil in a culture of technical rationality. *Public Integrity, 13*(3), 275–286. https://doi.org/10.2753/PIN1099-9922130307

Almeida, T., Abreu, F., & Ramalho, N. C. (2021). Becoming morally disengaged: How long does it take? *Leadership and Organization Development Journal, 42*(4), 548–563. https://doi.org/10.1108/LODJ-01-2020-0005

Ashforth, B. E., & Anand, V. (2003). The normalization of corruption in organizations. *Research in Organizational Behavior, 25*, 1–52. https://doi.org/10.1016/S0191-3085(03)25001-2

Bandura, A., Caprara, G. V., & Zsolnai, L. (2000). Corporate transgressions through moral disengagement. *Journal of Human Values, 6*(1), 57–64. https://doi.org/10.1177/097168580000600106

Barnes, J. A. (1994). *A pack of lies: Towards a sociology of lying*. Cambridge University Press.

Bird, F. B., & Waters, J. A. (1989). The moral muteness of managers. *California Management Review, 32*(1), 73–88. https://doi.org/10.2307/41166735

Dillard, J. F., & Ruchala, L. (2005). The rules are no game: From instrumental rationality to administrative evil. *Accounting, Auditing & Accountability Journal, 18*(5), 608–630. https://doi.org/10.1108/09513570510620475

Fleming, P., & Zyglidopoulos, S. C. (2008). The escalation of deception in organizations. *Journal of Business Ethics, 81*(4), 837–850. http://www.jstor.org/stable/25482258

Grauer, M. (2019). Ventilated rainscreen cladding system subframe contribution to annual source energy use in mid-size office buildings. *Energy and Buildings, 187*, 269–280. https://doi.org/10.1016/j.enbuild.2019.01.010

Grenfell Tower Inquiry. (2024a). *Grenfell tower inquiry: Phase 2 report overview*. UK Government. https://ctif.org/sites/default/files/2024-09/CCS0923434692-004_GTI%20Phase%202_Report%20Overview_E-Laying_0.pdf

Grenfell Tower Inquiry. (2024b). *Phase 1 report*. The National Archives. Retrieved October 20, 2025, from https://webarchive.nationalarchives.gov.uk/ukgwa/20250320040117/https://www.grenfelltowerinquiry.org.uk/phase-1-report

Grenfell Tower Inquiry. (2024c). *Phase 2 report*. The National Archives. Retrieved October 20, 2025, from https://webarchive.nationalarchives.gov.uk/ukgwa/20250319155650/https://www.grenfelltowerinquiry.org.uk/

Grenfell Tower Inquiry. (2024d). *Volume 1: Part 1 – introduction. Grenfell Tower Inquiry: Phase 2 report of the public inquiry into the fire at Grenfell Tower on 14 June 2017* (Vol. 1, Part 1). UK Government. https://webarchive.nationalarchives.gov.uk/ukgwa/20250320200657mp_/https://www.grenfelltowerinquiry.org.uk/sites/default/files/CCS0923434692-004_GTI%20Phase%202%20Volume%201_BOOKMARKED_0.pdf

Grenfell Tower Inquiry. (2024e). *Volume 2: Part 3 – The testing and marketing of key products. Grenfell Tower Inquiry: Phase 2 report of the public inquiry into the fire at Grenfell Tower on 14 June 2017* (Vol. 2, Part 3). UK Government. https://webarchive.nationalarchives.gov.uk/ukgwa/20250320200219mp_/https://www.grenfelltowerinquiry.org.uk/sites/default/files/CCS0923434692-004_GTI%20Phase%202%20Volume%202_BOOKMARKED_0.pdf

Grenfell Tower Inquiry. (2024f). *Volume 4: Part 6 – The refurbishment of Grenfell Tower. Grenfell Tower Inquiry: Phase 2 report of the public inquiry into the fire at Grenfell Tower on 14 June 2017* (Vol. 4, Part 6). UK Government. https://assets.publishing.service.gov.uk/media/66d817d9c52d5fb4c82dddf0/CCS0923434692-004_GTI_Phase_2_Volume_4_BOOKMARKED.pdf

Grover, S. L. (2005). The truth, the whole truth, and nothing but the truth: The causes and management of workplace lying. *Academy of Management Perspectives, 19*(2), 148–157. https://doi.org/10.5465/ame.2005.16965277

Hackitt, J. (2018). *Building a safer future independent review of building regulations and fire safety: Final report.* https://assets.publishing.service.gov.uk/government/uploads/system/uploads/attachment_data/file/707785/Building_a_Safer_Future_-_web.pdf

Hayes, J., Maslen, S., Holdsworth, S., & Sandri, O. (2021). Defining the capable engineer: Non-technical skills that support safe decisions in uncertain, dynamic situations. *Safety Science, 141.* https://doi.org/10.1016/j.ssci.2021.105324

Hayes, J., Maslen, S., & Schulman, P. (2024). A case of collective lying: How deceit becomes entrenched in organizational safety behavior. *Safety Science, 176*, 106554. https://doi.org/10.1016/j.ssci.2024.106554

Hodgson, J., & Balmer, A. (2022). Lying and time: Moving beyond the moral question of lying. *Sociology, 56*(5), 983–997. https://doi.org/10.1177/00380385211073233

Jenkins, S., & Delbridge, R. (2017). Trusted to deceive: A case study of 'strategic deception' and the normalization of lying at work. *Organization Studies, 38*(1), 53–76. https://doi.org/10.1177/0170840616655481

Jenkins, S., & Delbridge, R. (2020). Exploring organizational deception: Organizational contexts, social relations and types of lying. *Organization Theory, 1*, 1–24. https://doi.org/10.1177/2631787720919436

Jordan, J. (2009). A social cognition framework for examining moral awareness in managers and academics. *Journal of Business Ethics, 84*(2), 237–258. https://doi.org/10.1007/s10551-008-9706-3

Landry, J. R. (2009, May 28–30). Analyzing the London Ambulance Service's computer aided despatch (LASCAD) failure as a case of administrative evil. In *Proceedings of the special interest group on management information system's*

47th annual conference on Computer personnel research (pp. 167–174). ACM, Limerick, Ireland. https://doi.org/10.1145/1542130.1542163

Lilly, J., Durr, D., Grogan, A., & Super, J. F. (2021). Wells Fargo: Administrative evil and the pressure to conform. *Business Horizons, 64*(5), 587–597. https://doi.org/10.1016/j.bushor.2021.02.028

Maslen, S., Hayes, J., Holdsworth, S., & Sandri, O. (2021). When ethics is a technical matter: Engineers' strategic appeal to ethical considerations in advocating for system integrity. *Science and Engineering Ethics, 27*(4), 46. https://doi.org/10.1007/s11948-021-00324-7

Middlehurst, R., & Kennie, T. (1997). Leading professionals: Towards new concepts of professionalism. In J. Broadbent, M. Dietrich, & J. Roberts (Eds.), *The end of the professions? The restructuring of professional work*. Routledge.

Pidgeon, N., & O'Leary, M. (2000). Man-made disasters: Why technology and organizations (sometimes) fail. *Safety Science, 34*, 15–30.

Shilling, C., & Mellor, P. A. (2015). For a sociology of deceit: Doubled identities, interested actions and situational logics of opportunity. *Sociology, 49*(4), 607–623. https://doi.org/10.1177/0038038514546661

van de Poel, I. (2011). The relation between forward-looking and backward-looking responsibility. In N. A. Vincent, I. van de Poel, & J. Hoven (Eds.), *Moral responsibility: Beyond free will and determinism*. Springer.

van de Poel, I., Fahlquist, J. N., Doorn, N., Zwart, S., & Royakkers, L. (2012). The problem of many hands: Climate change as an example. *Science and Engineering Ethics, 18*, 49–67. https://doi.org/10.1007/s11948-011-9276-0

Virtanen, P., Lehtonen, T., & Raisio, H. (2022). Reaching into the dark side of organisations: The banality and emergence of administrative evil in the light of two case examples. *Scandinavian Journal of Public Administration, 26*(2), 3–21. https://doi.org/10.58235/sjpa.v26i2.7015

CHAPTER 3

The Invisible Influence of Expert Consultants in Engineering Risk-Based Decision Making

Jan Hayes

1 Introduction

How safe is safe enough? The accident record of major disasters illustrates graphically the potential for hazardous industries to kill people. Public safety and public trust are typical core values for engineers (Maslen et al., 2021; Trevelyan, 2014), and so preventing such accidents is a key ethical consideration for them. Despite this, high stakes engineering decision making is more usually framed by legal and technical considerations with questions of ethics pushed to the background if they are acknowledged at all. In the UK and Australia, "safe enough" is determined in hazardous industries by reference to the legal concept that risk must be reduced to a level that is as low as reasonably practicable (ALARP). The concept requires that all efforts should be made to reduce risk until the cost of further safety improvements is grossly disproportionate to the safety benefit gained. This general principle has led to the development of a range of practical engineering methods to support decision making ranging from

J. Hayes (✉)
School of Property, Construction and Project Management, RMIT University, Melbourne, VIC, Australia
e-mail: jan.hayes2@rmit.edu.au

J. Hayes, S. Maslen (eds.), *Ethics in the doing of technoscience*,
https://doi.org/10.1007/978-3-032-20002-0_3

quantitative assessments in the form of cost benefit analysis (CBA) to relatively unstructured expert judgments that enough has been done.

The safety literature is replete with case studies regarding application of the ALARP principle, and yet the ethical context within which the principle is applied has been little studied. I investigate how ALARP decision making is carried out in practice by drawing on a case study of the Australian pipeline engineering sector. I show that decisions on major safety expenditure involve three sets of actors: expert risk consultants, operating company personnel and regulators. Analysis of the empirical data is grounded in discourses on legal endogeneity, which address how abstract legal principles are enacted in practice and the associated interactions between regulators and those who are regulated. The chapter concludes that rather than being driven by questions of legal compliance, ALARP decision making is driven significantly by the professional value of public safety held by expert consultants and further, that this is little acknowledged in the industry.

2 Background

2.1 ALARP and the Law

The idea that employers have an ethical duty to do all that is "reasonably practicable" to ensure the safety of workers and the public has a long history (Ale, 2005; Hartford, 2009; Matthews & Ageros, 2010). To understand what ALARP means now, it is useful to understand the legal origins of the concept and how current thinking has developed from that. There is no simple definition of the legal expression "reasonably practicable". The starting point for discussion is two mid-twentieth-century judgments that addressed this issue. Legal interpretations in the UK are often traced to the definition by Asquith LJ in Edwards v National Coal Board (1949), quoted in Foster & Meredith (2025):

> "Reasonably practicable" ... seems to me to imply that a computation must be made by the owner, in which the quantum of risk is placed on one scale and the sacrifice involved in the measures necessary for averting the risk (whether in money, time or trouble) is placed in the other; and if it be shown that there is a gross disproportion between them—the risk being insignifi-

> cant in relation to the sacrifice—the defendants discharge the onus on them. (p. 537)

This definition of "reasonably practicable" was further considered in the House of Lords in Marshall v Gotham Co Ltd. in 1954 (Jones-Lee & Aven, 2011) and they found:

> The test of what is "reasonably practicable" is not simply what is practicable as a matter of engineering, but depends on the consideration, in the light of the whole circumstances at the time of the accident, whether the time, trouble and expense of the precautions suggested are or are not disproportionate to the risk involved, and also an assessment of the degree of security which the measures suggested may be expected to afford. (p. 877)

These two decisions are the origin of the principle that reducing risk to a level that is as low as reasonably practicable (ALARP) requires implementation of all measures that are practicable in engineering terms, unless the cost of implementing a given measure is grossly disproportionate to the benefit gained. As such, the work of deciding whether the risk presented by given facilities can be said to be ALARP falls to professional engineers, even when the ultimate decision making regarding additional risk reduction measures lies with senior management, depending on what resources may need to be expended.

In many jurisdictions, safety regulations are now goal-based and founded on internal control principles (Baram & Lindoe, 2014; Power, 1997). It is up to the company to decide how to control risks and achieve safety, provided it can be demonstrated that risk is controlled sufficiently. The need to make such a demonstration is where the ALARP principle comes to the forefront. Application of the ALARP principle requires weighing risk reduction versus the time, effort, and cost required to achieve it and so deciding whether a given risk reduction measure is justified (Hurst et al., 2019). Analysts must also pay attention to the "gross disproportion" idea so the weighing up is not a simple one-to-one assessment and is ultimately a matter of expert judgment. Thus, a matter of legal compliance has moved to a question of engineering judgment grounded in analysis but ultimately resting on questions of ethics.

This shift has not always been welcomed in technological bureaucracies seeking legal certainty. Operationalizing ALARP in the design of engineered systems was initially difficult for many companies as the principle was seen as "vague and unhelpful" (Kuo, 2001). Those responsible for compliance asked for more legal guidance, but this has generally not been forthcoming. In the legal literature, Ladkin (2019) addresses practical use of the functional safety standard IEC 61508 and how this links to the legal requirement of a duty of care in the UK. He notes that while engineers have written procedures to help them apply the ALARP principle, this is potentially misleading because "the arbiter of ALARP is not some engineering assessor but an English court of law. ALARP is a constraint on safety-critical system engineering, but it is not an engineering-scientific procedure" (p. 55). Thus, the ownership of the principle in practice has moved from law to engineering even though the final arbiter when it comes to compliance remains the legal system.

The literature on legal endogeneity is relevant here as it seeks to explain how companies interact with regulations and other legal institutions "to create ritualized ideas about what constitutes a rational response to law" (Edelman et al., 1999). Further, it is these "accounts, stories, and myths" about past interactions and what constitutes compliance with the law that ultimately determines the strategic responses of both companies and regulators, rather than the formal statute. In this way, the law becomes built in, or endogenous, to business practices (or in our case engineering practices).

Legal endogeneity theory sets out how legal requirements and company processes become merged. Firstly, when new legislative requirements are enacted, professionals external to both the regulated organization and the regulator are engaged to assist in the managerialization of the legal requirements. Studies of legal endogeneity have highlighted the role played by external professionals such as business consultants (Edelman et al., 1999; Gilad, 2014). These external professionals can construct a risk of legal non-compliance that they then fill by providing their expert services. Thus, organizations embed a version of the legal requirements—based on external advice of the meaning of compliance—into day-to-day business practices and systems. Compliance then becomes a matter of adhering to internal company systems. In a second phase, law becomes completely endogenous when a company's practices become accepted by the regulator as a proxy for compliance.

2.2 ALARP in Practice

The safety literature describes generically how the principle of ALARP should be put into practice. As the first step in applying the principle in any given situation, each identified risk is commonly categorized into one of three broad areas (Hurst et al., 2019).

1. Unacceptable: Risks that cannot be justified except in extraordinary circumstances. Risk reduction is required regardless of cost.
2. Tolerable: Risks that are tolerable in nature, but analysis is required to ensure that risk is reduced until the cost of further risk reduction is grossly disproportionate to the benefit gained.
3. Broadly Acceptable: Risks that are so low that further expenditure is not justified, and so existing risk reduction measures must be maintained, but are sufficient.

Based on these three broad categories, detailed analysis of risk reduction options to drive risk to ALARP then focuses on risks that are judged to fall into the second area. This has sometimes been called "the ALARP region" for short, and combined with confusion over the meaning of tolerability leads some practitioners to understand that risks falling into this zone are, by their nature, ALARP (Hurst et al., 2019). This is not the intention. In fact, tolerable risks are only acceptable in this scheme "once it has been demonstrated that all reasonably practicable risk reduction measures have been implemented" (Kannan et al., 2020).

Having narrowed the scope of which risks must be subject to detailed study regarding risk reduction, the question still remains as to how such demonstrations should be made. Consistent with the trend toward use of quantitative risk assessment (QRA) in the 1990s and early 2000s, in practice, demonstration that risk was ALARP often relied heavily on quantitative forms of cost benefit analysis (CBA) (Lindoe et al., 2014). Ultimately, CBA as a primary driver for decision making has proved to be both limiting and limited in application (Aven & Abrahamsen, 2007; French et al., 2005; Paterson, 2014). Rather than ALARP demonstration being a calculation with a deterministic outcome, Hurst et al. (2019) speak of "ALARP thinking" as telling a story that takes decision makers on a journey through the lifecycle stages of a major project from design, through operations, and onto decommissioning. Several researchers have proposed that multi-attribute methods are more comprehensive than CBA alone (French et al., 2005; Kletz, 2005; Li et al., 2024; Selvik et al., 2020) as their explicitly subjective nature forms a counter to the "illusory objectivity" of CBA (French et al., 2005). In most jurisdictions, CBA is now seen as something that may provide an input to decision making but not something that dictates an outcome.

Importantly, ALARP draws attention to risk reduction measures (rather than just risk per se) and guides engineers to take a cautionary approach where identified risk reduction measures should be implemented unless they can be shown to be not justified (Aven, 2009). Part of that decision making also includes consideration of whether the cost is grossly disproportionate to the safety benefit gained. However demonstrations of ALARP are made in practice, the principle is still very much alive in discussions of risk tolerability criteria and risk-based decision making. Recent literature demonstrates its use in a broad range of fields including nuclear power (Hurst et al., 2019), road design (Ridwan & Ayob, 2024), medicine (Patterson et al., 2024), shipbuilding (Li et al., 2024), fuel storage (Juwari et al., 2024), pesticide use (How et al., 2017) pipelines (Pei et al., 2018) and fire design of structures (Chaudhary et al., 2024). Interestingly, these studies originate from countries with a wide range of legal systems, many of which do not explicitly require compliance with the concept of ALARP.

As this brief review of the literature on the origins and current enactment of the ALARP principle shows, engagement with the principle has moved well beyond considerations of compliance with relevant legislation. What started as an obscure piece of common law in the UK now has global reach and has become endogenous to technical business processes that aim to ensure complex sociotechnical systems are safe enough.

None of these approaches explicitly link application of ALARP to the ethical roots of the issue it seeks to address, i.e., how safe is safe enough, although they are fundamentally utilitarian in nature. This school of ethical thought proposed by Jeremy Bentham in eighteenth-century England and further developed by John Stuart Mill is based on the idea that ethical actions are those that produce the greatest good for the greatest number. Further, the benefit of an action, its "goodness", is the pleasure or happiness it brings. Having said that, many utilitarian philosophers would find the idea of placing a monetary value on all benefits (as in CBA) to be problematic. This research aims to investigate how ALARP has been implemented in practice in one sector and whether/how the ethical aspect of risk decision making has been addressed.

3 Risk Management in the Australian Pipeline Sector

Australia's publicly owned natural gas transmission and distribution assets were privatized in the late twentieth century and the networks have been expanded under private ownership, now comprising around 42,000 km of transmission pipelines (APGA, n.d.). Most of the network is owned and operated by four major utility companies who supply gas to domestic households and commercial users. Other pipelines transport gas for export or to specific industrial customers. The overall public safety record of the sector is excellent with pipeline failure rates significantly lower than in other countries (Tuft & Cunha, 2013) although the potential for a serious accident remains. There have been a small number of significant pipeline failures (ABC, 2015; Bills & Agostini, 2009), but these have always occurred in remote areas. No member of the public has died as a result of gas pipeline infrastructure.

In Australia, regulation of onshore petroleum pipelines is a state matter. Legislation varies somewhat between the states, but there are two relevant factors in common. All states require that safety risk posed by pipelines is reduced as low as is reasonably practicable and all states require pipeline licensees to comply with the requirements of the relevant Australian Standard AS/NZS 2885. This Standard includes normative and informative material on how to determine whether risk is ALARP and so this material effectively had the weight of legislation. The Standard has a high degree of legitimacy with users, driven by its institutional links and the professional reputations of those responsible for the content (Hayes et al., 2022). This link between the legislation and the Standard is the beginning of the process whereby the exogenous requirements of the law become endogenous among companies, risk professionals and legal institutions, and ethical questions regarding public safety are operationalized.

4 Method

In the pipeline sector, application of ALARP to a given case is a professional activity that is temporally and spatially distributed. Some parts of the process are completed by individual engineers whereas others are group activities. There is significant uncertainty both in the details of how the process is applied and in the input data used to make decisions on outcomes. As such, qualitative methods are needed to understand how

ALARP is applied and the best methods of data collection are interview, supplemented by document review (Silverman, 2001).

Target interviewees were experts in pipeline sector risk management processes. Interview participants were recruited by direct invitation to individuals known to the research team as a result of over a decade of previous research with this sector. Twenty-five participants agreed to be interviewed including at least one risk practitioner from each major pipeline operating company and all the expert consultants who are used regularly by these companies to chair risk-related workshops. Regulators from four of the five states who regularly review company ALARP submissions also agreed to participate. Interviews were undertaken by video call. Interviewees were mostly men. All hold engineering qualifications and have over 20 years' experience in the gas pipeline industry.

Interviews were conducted in a semi-structured way with the conversation significantly driven by the responses of interviewees. Research ethics approval was obtained from the relevant university body. Audio recordings of the interviews were professionally transcribed for analysis. The data analysis involved coding toward a thematic analysis with a particular focus on discussions of ALARP in practice and orientation toward risk, ethics and ALARP. All interviewees were allocated a code consisting of a letter designating which type of organization employs them (R = regulator, C = consultant, P = pipeline company) and a unique number.

5 Findings

5.1 *Methods for Demonstrating ALARP*

Expert engineers adopt three different methods for reaching a determination that the risk posed by a given threat to pipeline integrity, and so public safety, is ALARP.

5.1.1 Deemed to Be ALARP

Safety decisions made during the pipeline design process are formally tested in a group setting known as a Safety Management Study (SMS) validation workshop. These workshops are run by an independent expert engineering facilitator and bring all engineers with a stake in the outcome together. This is the mechanism that can classify residual risk as intermediate, thereby triggering a requirement for further analysis to ensure the risk is tolerable. As one regulatory interviewee explained, getting an appropriate outcome is dependent on having the right people in the room with the right attitude: "Pretty easy to talk yourself out of things being credible. Pretty easy to talk yourself out of doing extra. You've really got to have the right people, the right mindset" (R03). Indeed sometimes, "there's a lot of push by certain people sometimes to make it low, not intermediate" (R03). Such a determination minimizes further work and further expenditure.

Once a risk has been found to be other than low, the intent of the standard is that the risk should be subject to further study to investigate risk reduction options, but some groups undertake ALARP assessment by consensus without further analysis. As one company interviewee explained:

> I think the Standard recommends a separate workshop. In my experience that's problematic for us … but we normally discuss the ALARP pretty much right there and then. And we generally try to get a consensus – as much as possible a consensus of whether the participants think that is ALARP or not. (P10)

No doubt getting a decision on the spot is expedient and some workshop facilitators are ready for this. Another consultant facilitator explained their practice, "I'm prepared in the workshop where it's obvious to everyone, there's agreement, consensus, even with my facilitation questions … that everyone's in agreement that it's clearly ALARP, then we'll record that as much in the workshop" (C13).

Some facilitators are prepared to facilitate ALARP determinations as part of SMS validation workshops, but others do not take on such assignments. One consultant described how they respond to such requests:

> And I say, well no we're not, because the whole point of the formal assessment, ALARP process, by token of the fact that you've got there, means you've got to do more work, and you've got to solve some difficult problems. And it's not something that we just deal with simply in a workshop as an add-on to the rest of the SMS workshop, it's an action that you guys have got to go and sort out afterwards. (C16)

Ultimately whether or not risk is ALARP is a matter of judgment. "It's not hard science" (R11), but the Standard encourages such judgments to be based on analysis as much as possible, rather than opinion and assertion. The problem is that "everybody just wants to get it done, and not actually do the real engineering to sort some of these problems out a bit" (C16). Without some additional engineering, decisions are not necessarily being made on an informed basis. As another facilitator explained, the value of deferring decision making until after further analysis is that it gives people "pause to realize by not taking action we're not at ALARP. They know deep down in their bones that they're not and they know that a detailed review will bear that out" (C18).

Deciding whether further risk reduction is justified in a workshop setting without further analysis puts pressure on the consultants who act as independent facilitators to become the ethical conscience of the group by challenging opinions expressed and the level of evidence available to support decision making. In this way they act as a counter to decisions that appear to be motivated by considerations of expediency, rather than doing the best for public safety.

5.1.2 ALARP Studies

The intention of the Standard is that formal ALARP assessments are engineering studies done based on data drawn from a wide range of sources. Many companies understand that and put significant effort into these studies which typically focus on understanding the likelihood of a pipeline failure due to a particular threat and the impact on that of a given risk reduction measure. As one interviewee explained, "I think that ALARPs tend to be done not in a workshop, they're written up by someone who puts it out for comment and discussion" (P15). As another interviewee pointed out, "a workshop is a terrible place to go into actual metadata" (P06).

This method combines both quantitative and qualitative aspects and requires judgment to determine the final outcomes based on assembled evidence. Technical work is done by experts who typically meet as a group with an independent facilitator to make the final decision as to whether more needs to be done for risk to be ALARP. In many cases, one individual has the task of conducting a detailed analysis, but consultation with multiple groups is required to collect relevant information:

> We need a lot of input from a lot of people. So, it is mainly engineering in terms of does it rupture? Does it leak? What's the failure event? What's the failure frequency? Chance of ignition, delayed ignition? So, we look into all that kind of stuff … we talk to [lands management] … and operations as well. What kind of size excavators do you expect to see out there? Do you

> see them with tiger teeth? ... we're building ... that metadata, so, at least we can back up our claims and say with confidence. (P06)

These studies are not trivial in the effort required. Even for relatively simple cases, "each ALARP assessment might take at least a month to finish" (C05). This work is often done by engineers working for operating companies but is also sometimes contracted out, typically to the same consultants who also facilitate SMS validation workshops. Part of their role is to critically challenge assertions made about control effectiveness and seek confirming data as one consultant explained:

> When people come up and they say, "Oh, we do patrols," [the facilitators] really try and probe them and work out how good they are. ... They think and probe, and really, do you really get out there once a week? Do you keep records? All these things. (C20)

Analysis methods can be both quantitative and qualitative. Quantitative methods have advantages:

> I think it's to give a little bit of structure around the comparable benefits of options. It's very easy to say, "This one's useful, that one's not useful", but with a semiquantitative method you can say that this one has this number, this one has that number, it's very clear what the difference is. So it helps to give visibility on that subjectivity of the usefulness of options. (P21)

But qualitative methods are useful too. As one interviewee explained, "we go through the list [of qualitative questions in the Standard], and we answer each question one by one" (C05).

A final meeting is often convened to discuss the results of the analysis with all parties and confirm a final conclusion. This is a reflection that there is still engineering judgment required to come to a firm conclusion when weighing up the pros and cons of a given risk reduction measure and deciding whether it should be implemented and/or whether further measures should be considered. As one consultant explained, "it's just it is subjective. I'm not sure if you can ever really get around that" (C22). The results can still be impacted by the attitudes of the people in the room and an open mind is important to get the best outcome:

> Really ideally, you've got to go into it with an open mind, the formal ALARP assessment, you don't really want to go into it if someone's already made

> their mind up. You don't want anyone going to it with any pre-conceived ideas. (C22)

Ultimately, it needs individuals to stand up for what they think is right as one consultant facilitator explained:

> It comes down to the mood in the room. It is—I reiterate to people when we're talking about ALARP it is a group decision, it is a consensus and that we have to—we have to address any dissenters. If a person doesn't feel that it's ALARP we have to back up and start all over again to go through the process—go through all the controls and really run it to ground. (C12)

5.1.3 Cost Benefit Analysis

Consistent with the broader literature as described above, most interviewees were of the view that CBA can support (but not dictate) the decision making process. As one consultant interviewee explained, "What I'm finding is … the industry is looking to move away from a purely CBA approach to focus more on the likelihood of failure and trying to get the likelihood of failure down as low as reasonably practicable" (C01).

This view was not universal. A small number of regulatory and operating company interviewees still yearn for an objective, deterministic method of making decisions. This regulatory interviewee was of the view that quantification of cost and benefit is the way to go and studies missing CBA are effectively incomplete:

> They might go into one part of it and discuss how much each different control would cost. But they may not quantify the risk reduction achieved by that control. And then relate that back to the cost. Sometimes they do and sometimes they don't. I think that's something that is often missing in ALARPs in general. (R08)

Another interviewee found use of qualitative methods to be ultimately unsatisfactory because they do not drive a firm conclusion. For this reason, they push for CBA within their company:

> I mean, they do come to a point where you just have to quantify it eventually because qualitatively, we all will come to a point that there is a risk … where we have to do something about it. It's a point where we have to come up with a cost benefit analysis. (P19)

Overall, most industry risk experts understand the limitations of CBA and would not rely on this method alone to drive outcomes. Nevertheless, some regulators and some companies prefer the illusion of accuracy, objectivity and certainty provided by numerical analysis.

5.2 *Benefits of Formal ALARP Demonstration*

It is one thing to conduct an engineering study that makes recommendations for safety expenditure, but it is another to get those in control of resources to commit to the necessary expenditure to make changes. Formal ALARP studies were seen by participants as valuable in the context of justifying decisions about necessary risk controls to those outside engineering such as senior management, external commercial stakeholders and regulators:

> The structure of the process really helps because you've put a lot of justification behind [a request for additional expenditure on risk control], ... it kind of really highlights to [senior management] that "hey, whoa, hang on a second, I need to pay attention to this". (P06)

Formal ALARP assessments are useful for communication with external stakeholders too,[1] thus serving as a resource in navigating ethical problems in practice:

> Outside the organization it has helped in understanding where—so, if we have to do something, we have to do something and it kind of just spells it out to them and they're just like, "Okay, we have to do that". So, it does put an end to that discussion, which is really good and yeah, it just comes down to cost and who paid for it. (P06)

Formal ALARP studies have been useful in justifying risk decisions to regulators since such studies provide "a formal justification for the company's position on a particular threat" (C17). These studies are useful because rather than focusing on risk as an abstract concept or providing a tick box for acceptability, they bring attention to action and questions of

[1] Since pipelines run through urban areas, sometimes the need for risk control needs to be negotiated with people outside the pipeline sector whose actions directly impact risk such as owners of other linear infrastructure whose work could physically damage a pipeline or property developers who may be building homes in adjacent land.

whether there are sufficient robust controls in place to protect the public from pipeline accidents. As another interviewee summarized, "effectiveness of controls is where it's at. How good is this control really? Let's get to the bottom of it. Is it going to work?" (C20).

6 Discussion

In the Australian pipeline industry, converting a legal requirement into an actionable process has moved the legal concept of ALARP into the realm of engineering decision making. Further, specialist consultants are a key source of expertise in putting processes for demonstrating that risk is ALARP into practice, i.e., endogenizing ALARP requirements. Other studies of legal endogeneity have noted the impact of business consultants and lawyers as professionals who "induce businesses to legalize themselves by socially constructing the risk of noncompliance with ambiguous laws" (Gilad, 2014, p. 137). We see the same broad influence in this case as the same small group of consultants each undertake other risk-related activities in the sector including running training regarding safety matters, conducting audits on behalf of government and participating in standards committees, and are thus seen as experts in effective and compliant risk management. As individuals, they are highly influential and act as a normative force within the industry, creating interpretations of the law that move from the consultants, to their clients and hence to the regulators.

Despite ALARP's origins in the law, legal compliance is not the consultants' primary motivating factor. In creating processes for demonstrating ALARP, the expert risk engineers acting as consultants construct "the meaning of compliance with ambiguous laws so that it matches their professional identities and enhances corporate demand for their services" (Gilad, 2014, p. 137). The endogeneity literature warns that in the translation into company systems and processes, legal requirements can be recast by professionals to embody managerial and commercial values, rather than legal compliance per se. We do not see that in this case. Rather, we see consultants' interpretations of legal requirements as being driven significantly by their professional ethics. This contributes to a high value placed on public safety in company decision making and so what occurs in practice.

Further, ALARP is widely understood in the literature to be ultimately a matter of professional judgment as both what is reasonable and what is practicable must be adapted to suit the specific context (Selvik et al.,

2020). In our case, we see that the expert consultants have embraced the subjective nature of ALARP decision making and see their role as facilitators of a process that ensures a broad range of relevant information is collected and considered by a group of experts who have a focus on considerations such as possible adverse outcomes, the likelihood of such outcomes and the cost of taking (further) action. A significant aspect of their role is to ensure that decisions are made by the right people with the right mindset considering the right information. This preferred method brings consideration closer to the practicalities of the threats to an operating pipeline and the disasters that these engineering choices aim to avoid, thereby putting conversations more in touch with the ethical questions involved. Decisions made in this manner are seen to be better than workshop-based opinion engineering consensus based on limited data or more abstract quantification in the form of CBA. Both options put decision making further away from the reality of field experience.

In contrast to the deep engagement of consultants, regulatory participants observe industry trends in relation to demonstration of ALARP, but they do not get involved. When regulators speak of how industry understanding has evolved and improved, they are speaking as observers, rather than active drivers of this process. Change seems to stem from the consultants who thus also "play an active part in the process of endogeneity by transmitting and injecting institutionalized business practices into legal jurisdictions" (Gilad, 2014, p. 137). As such, both the law and the regulators act in the background.

Such a process is open to manipulation and so the consultant's work is not completely without conflict as the data shows. Both consultants and regulators report variation between pipeline companies when it comes to their level of maturity regarding risk-based decision making. Some companies apparently prefer less robust and so arguably less ethical decision making approaches seemingly driven by time and cost considerations. Consultants describe how they guide clients who want to manipulate results or take what they perceive as short cuts by trying to find a middle way between what they believe is the best method and what the company prefers to do. They pick their battles by encouraging further analysis on cases that they believe most require it. Their approach is action-oriented, promoting what they believe to be ethical outcomes but not usually expressed in ethical terms (Maslen et al., 2021).

7 Implications

The ALARP principle has significant currency in the Australian pipeline sector largely through the actions of expert consultants. When it comes to legal compliance, the various approaches taken have been accepted by regulators as meeting regulatory requirements for approvals and licensing. The ultimate test would be in the courts, but prosecutions under pipeline legislation are rare. Unless and until ALARP is tested in the courts, the sector has largely handed responsibility for defining how risk decisions are made to a group of very experienced risk experts who serve the industry as consultants. This finding is not unique to this case. Størkersen et al. (2023) found that specialist risk consultants in the Norwegian offshore oil and gas sector play a major role in the riskwork (Power, 2016) of operating companies seeing themselves as in the middle of the operating company/regulator dyad. The Australian risk experts see their role somewhat differently. The legislated requirements and the regulators are less in focus and the consultants seem to be driven directly by professional values of regard for public safety.

It is interesting to consider why the critical role of consultants and their professional values might be so apparently invisible. One explanation is the common framing of risk regulation in accordance with the internal control principle (Baram & Lindoe, 2014), i.e., regulation is based on the assumption that risks are knowable and controllable. In this framing of risk, methods are established and drive results of risk studies in deterministic ways, so consultants are viewed as expert resources applying well-defined methods in a very narrow way to specific cases with control residing largely with the company procedure or industry guidance. Our case study is a demonstration that this is not the case. Behind this veneer of objectivity and determinism projected by many engineering discussions of ALARP, the messy reality is of significant variation in ALARP in practice even in one industry and one location which belies the impression of a common risk management framework. As Clarke highlighted several decades ago, "risk analyses are *social* products and therefore open to many sources of bias and distortion" (Clarke, 1988, p. 155).

Previous research has noted that the role of expert consultants in accident prevention is both under examined and under theorized (Størkersen et al., 2023). Accordingly, this study indicates that there would be much value in further exploring the role of consultants in influencing corporate safety decision making and operationalization of ALARP in a range of safety domains. Such studies should seek to go beyond the specific technical methods adopted in pursuit of ALARP such as CBA and should more broadly engage with the ethical, social and organizational work that is done regarding what is considered in reaching a conclusion and whose views are heard. The response of regulators is another important dimension.

Acknowledgments This work is funded by the Future Fuels CRC, supported through the Australian Government's Cooperative Research Centres Program. The cash and in-kind support from the industry participants is also gratefully acknowledged.

References

ABC. (2015). Gas pipeline rupture in South Australia 'may take days to repair' leaving thousands without gas. *ABC News*. https://www.abc.net.au/news/2015-04-14/cold-showers-for-sa-residents-gas-outage/6390516

Ale, B. J. M. (2005). Tolerable or acceptable: A comparison of risk regulation in the United Kingdom and the Netherlands. *Risk Analysis, 25*(2), 231–241.

APGA. (n.d.). *Pipeline facts and figures*. Australian Pipelines and Gas Association. Retrieved September 9, 2024, from https://apga.org.au/pipeline-facts-and-figures

Aven, T. (2009). Perspectives on risk in a decision-making context - Review and discussion. *Safety Science, 47*, 798–806.

Aven, T., & Abrahamsen, E. (2007). On the use of cost-benefit analysis in ALARP processes. *International Journal of Performability Engineering, 3*(3), 345–353.

Baram, M., & Lindoe, P. (2014). Modes of risk regulation for prevention of major industrial accidents. In M. Baram, P. Lindoe, & O. Renn (Eds.), *Governing risk in offshore oil and gas operations* (pp. 34–55). Cambridge University Press.

Bills, K., & Agostini, D. (2009). *Offshore petroleum safety regulation; Varanus island incident investigation*. Government of Western Australia.

Chaudhary, R. K., Gernay, T., & Van Coile, R. (2024). Multi-objective optimization of structural fire design. *Fire Safety Journal, 146*, 104139. https://doi.org/10.1016/j.firesaf.2024.104139

Clarke, L. (1988). Politics and bias in risk assessment. *The Social Science Journal, 25*(2), 155–165. https://doi.org/10.1016/0362-3319(88)90003-1

Edelman, L. B., Uggen, C., & Erlanger, H. S. (1999). The endogeneity of legal regulation: Grievance procedures as rational myth. *American Journal of Sociology, 105*(2), 406–454. https://doi.org/10.1086/210316

Foster, N., & Meredith, J. (2025). *Workplace Health and Safety Law in Australia* (Third ed.). LexisNexis.

French, S., Bedford, T., & Atherton, E. (2005). Supporting ALARP decision making by cost benefit analysis and multiattribute utility theory. *Journal of Risk Research, 8*(3), 207–223. https://doi.org/10.1080/1366987042000192408

Gilad, S. (2014). Beyond endogeneity: How firms and regulators co-construct the meaning of regulation. *Law & Policy, 36*(2), 134–164. https://doi.org/10.1111/lapo.12017

Hartford, D. (2009). Legal framework considerations in the development of risk acceptance criteria. *Structural Safety, 31*, 118–123.

Hayes, J., Maslen, S., & Merad, M. (2022). The making of a legitimate standard. *Safety Science, 152*, 105771. https://doi.org/10.1016/j.ssci.2022.105771

How, V., Abdullah, K., & Othman, K. B. (2017). Evaluate the decision of as low as reasonably practicable (ALARP) solutions for the safe use of pesticide among the pesticide handlers, paddy farm, Tanjung Karang, Selangor. *Annals of Tropical Medicine and Public Health, 10*(1), 36–43. https://doi.org/10.4103/ATMPH.ATMPH_73_17

Hurst, J., McIntyre, J., Tamauchi, Y., Kinuhata, H., & Kodama, T. (2019). A summary of the 'ALARP' principle and associated thinking. *Journal of Nuclear Science and Technology, 56*(2), 241–253. https://doi.org/10.1080/00223131.2018.1551814

Jones-Lee, M., & Aven, T. (2011). ALARP—What does it really mean? *Reliability Engineering and System Safety, 96*, 877–882.

Juwari, J., Anugraha, R. P., Zanata, M., Leksono, N. A., Hamid, M. D., & Cahyono, T. (2024). Quantitative risk assessment and risk reduction of worst-case accident scenario at fuel storage terminal. *Journal of Loss Prevention in the Process Industries, 88*, 105272. https://doi.org/10.1016/j.jlp.2024.105272

Kannan, S. K., Guinchard, F., Matsui, T., & Takahashi, Y. (2020). As low as reasonably practicable (Alarp) demonstration: A case study on firewater curtain application for liquified petroleum gas(lpg) sphere. *Pertanika Journal of Science and Technology, 28*(Special Issue 1), 99–120.

Kletz, T. A. (2005). Looking beyond ALARP: Overcoming its limitations. *Process Safety and Environmental Protection, 83*(2), 81–84.

Kuo, C. (2001). Safety of offshore installations – Making ALARP principle more practicable. In *ISOPE International ocean and polar engineering conference*. ISOPE.

Ladkin, P. B. (2019). Duty of care and engineering functional-safety standards. *Digital Evidence and Electronic Signature Law Review, 16*, 51–56.

Li, X., Zhang, X., & Yuan, Y. (2024). Dynamic multi-attribute decision-making method for risk-based ship design. *Applied Sciences (Switzerland), 14*(13), 5387. https://doi.org/10.3390/app14135387

Lindoe, P. H., Baram, M., & Renn, O. (Eds.). (2014). *Risk governance of offshore oil and gas operations*. Cambridge University Press.

Maslen, S., Hayes, J., Holdsworth, S., & Sandri, O. (2021). When ethics is a technical matter: Engineers' strategic appeal to ethical considerations in advocating for system integrity. *Science and Engineering Ethics, 27*(4). https://doi.org/10.1007/s11948-021-00324-7

Matthews, R., & Ageros, J. (2010). *Health and safety enforcement: Law and practice.* Oxford University Press.

Paterson, J. (2014). Safety regulation on the UK continental shelf. In M. Baram, P. Lindoe, & O. Renn (Eds.), *Governing risk in offshore oil and gas operations.* Cambridge University Press.

Patterson, M., Campbell-Adams, T., & Johnson, B. (2024). Perception of patient habitus and the influence on exposure factor selection and resultant imparted dose. *Radiography, 30*(3), 793–798. https://doi.org/10.1016/j.radi.2024.02.016

Pei, J., Wang, G., Luo, S., & Luo, Y. (2018). Societal risk acceptance criteria for pressure pipelines in China. *Safety Science, 109*, 20–26. https://doi.org/10.1016/j.ssci.2018.05.006

Power, M. (1997). *The audit society: Rituals of verification.* Oxford University Press.

Power, M. (Ed.). (2016). *Riskwork: The organizational life of risk management.* Oxford University Press.

Ridwan, T. M., & Ayob, A. F. (2024). Risk management of coastal ring roads to rate abrasion on the Lhokseumawe City. *Journal of Advanced Research in Applied Sciences and Engineering Technology, 39*(1), 168–182. https://doi.org/10.37934/araset.39.1.168182

Selvik, J. T., Elvik, R., & Abrahamsen, E. B. (2020). Can the use of road safety measures on national roads in Norway be interpreted as an informal application of the ALARP principle? *Accident Analysis and Prevention, 135*, 105363. https://doi.org/10.1016/j.aap.2019.105363

Silverman, D. (2001). *Interpreting qualitative data: Methods for analysing text, talk and interaction.* Sage.

Størkersen, K., Hayes, J., Standal, M. I., Ognedal, M., & Skogstad, M. R. (2023). We're in the middle of it: Consultants' role in risk management in the Norwegian petroleum sector. *Safety Science, 167*, 106258. https://doi.org/10.1016/j.ssci.2023.106258

Trevelyan, J. (2014). *The making of an expert engineer.* CRC Press.

Tuft, P., & Cunha, S. (2013). Comparing international pipeline failure rates. *Journal of Pipeline Engineering, 4th Quarter*, 313–319.

CHAPTER 4

Industrial Development, Invisible Hazards, and Bringing to Light Their Entanglements: Lessons for Ethics from the 2014 Kaohsiung Propene Explosions in Taiwan

Bono Po-Jen Shih, Wen-Ling Hong, and Jr-Ping Wang

1 Introduction

Engineers have argued that we benefit more from failure than success in advancing our knowledge and tools for building safer technological systems (e.g., Petroski, 1985). However, such a premise is often not viable without robust democracy, which recognizes civil discourse, public oversight, and government accountability as cornerstones of advancing governance and policy to support the development and application of professional guidelines and ethics. In the West, the 1960s saw the publication of Rachel Carson's *Silent Spring* and Ralph Nader's *Unsafe at Any*

B. P.-J. Shih (✉)
Pennsylvania State University, Pennsylvania, PA, USA
e-mail: bonoshi@vt.edu

W.-L. Hong • J.-P. Wang
National Kaohsiung University of Science and Technology, Kaohsiung, Taiwan
e-mail: wlhong@nkust.edu.tw; jpw@nkust.edu.tw

J. Hayes, S. Maslen (eds.), *Ethics in the doing of technoscience*,
https://doi.org/10.1007/978-3-032-20002-0_4

Speed, which generated public pressure and citizen activism that shaped government policy and corporate liability for ethical development of technology. Their works specifically inspired and provided a foundation for the emerging field of science, technology, and society (STS) in the late 1960s and engineering ethics in the mid-1970s (Cutcliffe, 2000; Herkert, 2005; Weil, 1984). In this context, Western engineering ethics has acknowledged the sway from governments, employers, and even other engineers as sources of potential conflicts and therefore strived to uphold individual engineers as "autonomous, reasonable, and responsible moral agents" (Martin & Schinzinger, 1983, pp. 6, 18).

This understanding and experience of engineering ethics is nonetheless not global. Before the lifting of martial law in 1987, there was little room in Taiwan for challenging the legitimacy of industrial development policies as well as the environmental impact of the industry and factories under the auspices of the ruling party Kuomintang's (KMT) regime. After 1987, although civil society began to scrutinize the nexus between industry and government and demand accountability for the harm resulting from unchecked development, systematic negligence and ignorance were already entrenched in society and even persisted into the twenty-first century. As will be demonstrated, institutional lack of awareness and understanding of industrial risk became prominent in the 2014 Kaohsiung propene explosions, causing 32 deaths, over 300 injuries, and the destruction of more than 4 kilometers of major roads.

The following section begins our case study of the Kaohsiung propene explosions by tracing the social history of engineering ethics in Taiwan, from the martial law regime, the democratizing era, to the recent engagement of the Taiwanese STS community. It then tells the story of the explosions through an STS lens, recounting the incident and immediate responses, prior pipeline operations, and judicial responses. We continue by examining the systematic ignorance of risk and the significance of governance reform and civic engagement for supporting ethical behavior. We conclude with a synthesis of key insights, ongoing developments, and broader implications.

2 The Social History of Engineering Ethics in Taiwan

2.1 Ethics Under Martial Law (Pre-1987)

Without the liberty and freedom for individuals to publicly express dissenting opinions or actions against the authorities, ethics in engineering in Taiwan was by and large a simplistic industrial "work ethic" (Shih & James, 2024, pp. 147–148), by which engineers committed to national development through the material provisioning of efficient and useful technology (Chang et al., 2025). In light of environmental issues and social disorder in industrial society, leading technocrats (e.g., Li, 1981) often pointed to the failure of individuals to extend their care and moral responsibilities—normatively reserved for relatives and friends according to traditional Confucian relational ethics—to anonymous, indefinite people. However, forward-thinking critics who (very implicitly) disagreed with this view argued that the development of individual ethics must come from democracy and the protection of "freedom of speech, thoughts, and publication" under the rule of law, rather than being dictated by paternalistic authority (Wei, 1981). The issue of inadequate ethics could not be attributed to any single individual but was a pervasive, social-structural problem that should be tackled first by government elites, who needed to abstain from bureaucratic and corrupt practices and focus on improving social, economic systems (Hsiao, 1981).

As a consequence, when early commentators used (and somewhat improvised) the collocation *gong cheng lun li* (工程倫理) or "engineering ethics" in Chinese, they often conceptualized "engineering" as a collective enterprise shaped largely by government departments, businesses, top officials, factory owners, and other decision makers, rather than based on the call for autonomy of individual engineers, who after all had limited agency and could not reasonably be accountable. While a similar lack of autonomy among engineers was also recognized in Japan, where some commentators provokingly asked if engineers "can … get out of [the] slavery" of corporate culture (Natsume, 2021, p. 162), in Taiwan, thinking of engineering as a collective is further amplified because "engineering," as understood in native Chinese, is distinctly polysemous and inclusive of technical work, as its meanings encompass: (1) building, bridges, and other structures that are large and complex and need careful planning and coordination; (2) large-scaled or complex technological activities that follow specific plans, such as construction, mining, water conservancy, power generation, and manufacturing; and the predominantly Western sense of engineering in terms of (3) the branch of science and technology concerning machines, structures, or other similarly complex systems or processes that need the specialized knowledge and skills of professionals (Shih, 2021).

Influenced by the strong focus on engineering in the collective sense, early adopters of "engineering ethics" during the late martial law developed a tradition of contemplating macroscopic ethical questions concerning engineering activities (e.g., Anonymous, 1986; Chang, 1987). The scope of engineering ethics, naturally, included business ethics, governance ethics, development (policy) ethics, and environmental ethics, along with the infrastructure needed to encourage and support ethical engineering practices such as through regulations and ethics education. During this era, however, engineering ethics in Taiwan concentrated on widespread, high-profile issues that had caused significant public outcry and sacrifice such as environmental degradation and the frequent corruption in construction. In its efforts to balance public goods of development with (presumably equally) shared environmental and societal costs, engineering ethics was idealistically constrained to a utilitarian paradigm of cost-benefit analysis.

2.2 Developments of Engineering Ethics After 1987

Against the above political backdrop, the lifting of martial law in 1987 anticipated extensive civic engagement in ethical questions and controversies in science and technology. In January 1991, a government-led national science and technology conference placed "ethical issues of science and technology development" at the top of a long, 13 item list of suggested new fundamental research areas in the humanities and social sciences (National Science Council, 1991). Specifically, participants acknowledged that social and cultural problems linked to Taiwan's rapid economic development in the past decades had not been studied with "new ideological trends" (新思潮) from the globe. Considering that democratized Taiwan needed public participation in the formulation and assessment of science and technology policies, conference organizers recommended providing courses like "science and technology ethics" and "engineering ethics" in colleges and utilizing mass media and other life-long education methods to enlighten citizens.

Recognizing that "the presence of a course on ethics in the undergraduate engineering curriculum is indicative of changes … in the society" (Wilcox, 1983, p. 15, cited in Wang, 1992), scholars who started teaching engineering ethics in the early 1990s were motivated by the paradox of freedom in the turbulent, transformative years of democratization. On the one hand, the increasing freedom of speech, association, and assembly made visible a surge of more concrete, lingering sociopolitical problems attributable to rapid industrialization and economic development in the past. To early participants, challenges and issues interfacing with engineers' professional work and responsibilities appeared overwhelmingly vast and more complicated than ever, justifying the urgent need for engineering ethics (Teng, 1991; Wang, 1994). On the other hand, the initial democratization of Taiwan was a bumpy ride, when simmering tension and dissatisfaction with the government burst into conspicuous political struggles and confrontations (Chao & Myers, 1998; Rigger, 1999). Contemporaries were worried that the politicization process had compromised public morality (Wang, 1992, p. 66) and professional responsibility (Pan, 1992; Teng, 1991, p. 319). As a remedy for the perceived moral decline and social disorder following the lifting of martial law, engineering ethics was expected to help inculcate professional values and social responsibility in the aspiring engineers (Teng, 1991; Wang, 1991).

As part of the contributions, early engineering ethics educators formalized the broad tradition of "engineering ethics," which included many macro-ethical issues not typically under the purview of individual engineers, such as copyrights, patents, genetic engineering, environmental and urban planning, the social impacts of automation, energy and ethics, information ethics, and data ethics. The inclusion of a wider array of topics in "engineering ethics" reflected, again, the polysemous nature of "engineering" in native Chinese terms (Shih, 2021). Engineering academics, rather than social scientists or historians, led the development of engineering ethics education. In the longer term, engineering professors and practitioners acquired substantial expertise and started publishing native engineering ethics textbooks in the late 2000s, moving away from the reliance on translating Western works (Tai & Hong, 2018). Building on the previous generation's tendency to broadly encompass technological activities in engineering ethics, several Taiwanese engineering ethics textbooks were intended for the general public, who should all be deemed as stakeholders in engineering (e.g., Chin & Chiang, 2017; Chou, 2014). Some also embraced "ethics of technology" in the book content (e.g., Yeh et al., 2017).

Whereas the engineering ethics movement starting in the 1990s was characterized by engineering scholars teaching from within engineering, the 2000s saw concerted, interdisciplinary efforts from the emerging Taiwanese science and technology studies (STS) community highlighting the social, political context of engineering practices (Tai & Hong, 2018). The lifting of martial law had not immediately aroused significant interest in Taiwan's contemporary technological issues among local historians and philosophers, who attended more to theoretical, philosophical questions about science. It was not until near 2000 that historians of science and technology "broke up" (Chu, 2010, p. 2) with philosophers, formed an STS academic community, and developed a Taiwanese STS journal. In the first two issues in 2001 and 2002, local scholars focused predominantly on the history of technological development and controversies in modern Taiwan.

Owing to the support from the Ministry of Education, the STS community thrived in the 2000s by growing an interdisciplinary network of scholars. The welcoming gesture of the community invited scientists and engineers to collaborate with scholars in the humanities and social sciences and engage in public discourse on technological issues. Some members also participated in high-profile controversies, including the legal proceedings against the RCA Corporation by its former workers (Chen, 2023), as well as the identification and monitoring of pollution from Formosa Plastics, the largest private petrochemical company in Taiwan.

Several interdisciplinary scholars started enriching engineering ethics with STS in the late 2000s (Tai & Hong, 2018). Collectively, they called into question the one-dimensional, technical mentalities in engineering and introduced perspectives from the social sciences and humanities, including the concepts of risk society, normalization of deviance, appropriate technology, the politics of technology, gendered and gendering technology, lay expertise, etc. Developing local ethics case studies based on current, ongoing technological controversies and recent engineering disasters, the STS approach to engineering ethics in Taiwan has uncovered the sociopolitical history of technological development that shaped engineering practice and systems, showing how sociopolitical processes and systems interact with engineering knowledge and work.

This experience of STS within engineering ethics presents an interesting historical contrast with the Western counterpart. Although Western scholars also recognized that STS would enhance engineers' sensitivity to workplace culture and other contexts that impact their decision making processes (Lynch & Kline, 2000), critics had argued that in engineering disasters where sociopolitical constraints severely limited engineers' decision space, the STS approach did not concern engineering ethics at all because "[d]ecision-makers are understood to have 'no choice' (hence, no [real] decision)" (Davis, 2006, p. 226). In comparison, engineering ethics in Taiwan extends beyond "engineers" to include "the engineering personnel" (工程人員), encompassing non-engineer actors who make decisions and impact engineering activities. As a result, tensions between rank-and-file engineers and "real" decision makers have theoretically been less of an issue, as they are expected to be internally resolved and reconciled in favor of a shared ethical stance (Shih, 2021).

3 The 2014 Kaohsiung Propene Explosions: An STS Account[1]

To illustrate what is at stake within the sociopolitical factors of engineering ethics, we provide our following account of the Kaohsiung propene explosions. Our narrative, shaped by our positionality as Taiwanese scholars in engineering ethics and STS (Hong & Wang, 2022; Hong et al., 2018), challenges the conventional framing of practitioners as heroic or tragic figures making an individual(istic), right-or-wrong decision at a critical juncture of a disaster. Instead, the landscape of public policy and civil discourse on technological development has mediated the microscopic ethical behavior of technical practitioners. We therefore invite readers to consider practitioners as citizens participating in, and being affected by, the sociopolitical processes of technological development that provide an infrastructure of (un)safety cultures that support (or impair) the ethical behaviors and decisions of these professionals.

3.1 The Incident and Multi-site Responses

On the surface, the propene explosions that struck Kaohsiung downtown areas—where population density exceeds 15,000 people/km^2—were due to a lack of experience responding to hazardous chemicals leakage on the day of July 31, 2014. Between 8:46 p.m. and 8:50 p.m., three hours before the explosions, citizens in downtown Kaohsiung began reporting

[1] The details and generally accepted facts about the incident in this section are based on court judgments available at https://judgment.judicial.gov.tw/.

to the Fire Department that they smelled "natural gas" and saw white smoke coming out of manhole covers around the intersection of Ersheng Road and Kaixuan Road. Around 9 p.m., the Fire Department implemented traffic control and contacted natural gas companies requesting pipeline information and field inspection. Some firefighters suspected that the leaked gas was not natural gas because of a distinctive, unfamiliar smell. An hour after arriving at the scene, the Fire Department determined that the leaked substance was not natural gas from the reply of the only natural gas supply company in the area. At 9:54 p.m., they started calling Chinese Petroleum Corporation (CPC Corporation)—the state-owned oil and gas company—for its pipeline information. CPC Corporation initially responded that it had "no (natural gas) pipeline" in the area, but they soon found out that there were three petrochemical pipelines there: one it owned and two belonging to two other companies—China Petrochemical Development Corporation and LCY Chemical Corporation (LCY Chemical)—carrying either propene or ethene. One CPC Corporation representative arrived at the site at 10:35 p.m. and informed the authorities what they knew. However, because of the chaos at the scene, the information was neglected and did not sink in among onsite personnel. Also, the city's pipeline mapping information system had a software design flaw (by its information service contractor) preventing the Public Works Bureau from identifying LCY Chemical as the owner of the leaky pipeline. Therefore, the Fire Department did not ask LCY Chemical about the operational status of its petrochemical pipelines, nor did they request its representatives to come to the scene to investigate and clarify the situation.

At 10:30 p.m., a chemical emergency response team also arrived. The leaking gas was dispersing along the ground and not behaving like natural gas, which rises in air. However, initial tests still focused on natural gas and confirmed that this was not the leaking substance. The response team was later informed that there were also ethene and propene pipelines under the ground, and by 11:35 p.m., they were able to detect the presence of these chemicals with portable instruments. Unfortunately, at 11:56 p.m., an unknown heat source ignited the leaking gas and caused a series of massive explosions of the culvert filled with dense gas underneath the roads. The explosions blew open and destroyed more than 4 kilometers of major roads and caused 32 deaths and over 300 injuries (Fig. 4.1). Many vehicles were flipped over and trapped in trenches, with some even sent to rooftops (Fig. 4.2). It was not until 6:30 a.m. of the next day that the leaked gas

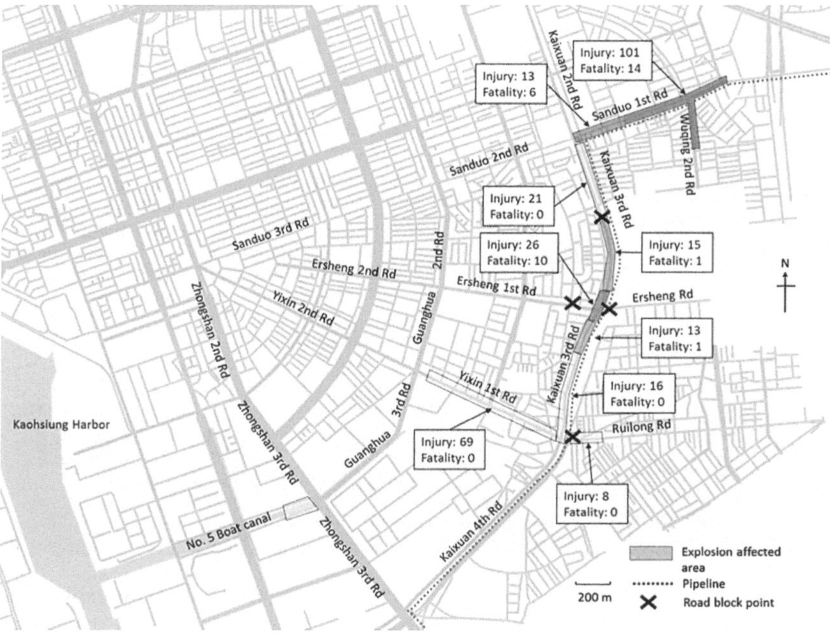

Fig. 4.1 Map of explosions and casualties. (Figure courtesy of Prof. Jenq-Renn Chen.)

was confirmed to be propene through lab spectral analysis (Yang et al., 2016).

In parallel during the critical time span, LCY Chemical expected to receive liquefied propene as feedstock to its inland factory through the pipeline it owned. Another party, China General Terminal & Distribution (CGTD) Corporation, was responsible for transporting the propene from storage tanks near the harbor area. At 8:50 p.m., an LCY Chemical operator found that the receiving flow rate of propene had dropped to zero and notified the foreperson. LCY Chemical called CGTD Corporation at 8:55 p.m., and the latter's representatives found that CGTD's supply pump was overloaded, outputting an excessive flow of propene but unable to maintain even half of the expected pipeline pressure. The pump was stopped, but the two parties did not perform an effective test for pipeline integrity. Believing that the pipeline was safe to operate, the two companies resumed transporting propene at 10:15 pm. Even though significantly

Fig. 4.2 Destruction of roads and vehicles resulting from the propene explosions in Kaohsiung, 2014. (Photo by Chih Nan Fu. Courtesy of Citizen of the Earth, Taiwan.)

less propene was arriving at the plant than being supplied and the pipeline pressure was far below normal, pumping continued for more than an hour despite these clear signs that something was wrong. At 11:23 pm, an off-duty CGTD foreperson commuting to work happened to pass by the area of concern and recognized the smell of propene. After arriving at the CGTD facility at 11:33 p.m., the foreperson instructed onsite operators to immediately shut off the pump. Nevertheless, the remedy was too late. A substantial amount of liquefied propene remained in the pipeline and continued vaporizing through the leak point, until the explosions at 11:56 p.m.

3.2 *Pipeline Operations Prior to the Explosions*

A more thorough investigation, however, would reveal that the incident was not simply due to a series of mishaps or incompetence on that day but also stemmed from negligence and ignorance spanning over the previous two decades. A section of the pipeline (4 inches in diameter) that should have been buried in the soil was instead exposed to the air within an underground drainage culvert, which compromised the pipeline's corrosion protection. Humidity caused the pipeline to corrode until it could no longer withstand the pressure and burst. Liquefied propene expanded into gas and quickly filled the culvert system, spreading widely underneath major roads before the explosions (Fig. 4.3).

The propene pipeline in question was built between 1990 and 1991 by CPC Corporation on behalf of Taiwan Polypropylene—the first polypropylene manufacturer in Taiwan. The pipeline contractor found out that a culvert that would soon be constructed by the city government would cross the pipeline route and therefore designed the intersected pipeline section to run above the planned culvert. In fact, a new route was chosen for the culvert when it was later constructed in 1992. The engineers in charge of supervising or inspecting the culvert construction did not inform the pipeline owner that its pipeline now ran through the upper part of the culvert, rather than entirely above it, even though expectations of such communication and collaboration were documented in the construction drawing and pre-construction meetings. LCY Chemical acquired the pipeline in 2008 from Taiwan Polypropylene, but it never inspected the corrosion protection of the pipeline, assuming that its pipeline would be covered by the corrosion protection arrangements for the two other parallel pipelines.

Fig. 4.3 Culvert (east-west) containing the 4-inch propene pipeline (north-south). (Image courtesy of Metal Industries Research & Development Centre, Taiwan. Annotations translated by the authors.)

3.3 *Judicial Understanding of Responsibility and No Responsibility*

Court rulings revealed how legal responsibility ultimately fell on each party. With regard to criminal liability, the Supreme Court of Taiwan ruled that the city government's three engineers who were responsible for supervising or inspecting the culvert construction were guilty of negligence leading to the death and injury of other persons while clearing all private company personnel of criminal punishment. In terms of civil liability, a higher court's ruling determined that the city government was responsible for most (80%) civil damages to the victims, not simply because of the negligence of government engineers. The Fire Department and other emergency response agencies assumed a natural gas leak, followed established procedures, and were not able to quickly identify propene despite Kaohsiung's long history of underground pipelines transporting hazardous petrochemicals. The emergency response teams did not establish an effective incident command system, resulting in chaos and confusion at the scene and delays in informing petrochemical testing teams about the potential substances. Moreover, the Public Works Bureau's pipeline mapping information system failed to display LCY Chemical's pipeline, and the city government neglected to validate the system for reliability and accuracy. In addition, the court ordered CGTD Corporation and LCY Chemical to share the remaining (20%) civil liabilities. Particularly, LCY Chemical should have regularly maintained and inspected the pipelines it owned according to industry standards and city ordinances. CGTD Corporation and LCY Chemical's on-duty personnel (including engineers, forepersons, and operators) did not follow their respective companies' emergency response guidelines and protocols, and their engineers also failed to conduct an effective pipeline integrity test during the leakage.

However, despite civil liabilities, court records show that judges were sympathetic toward the private company personnel in many instances, foregrounding the participants' limited knowledge about the petrochemical pipelines that were entangled with their daily practice. For example, in criminal judgments, the Supreme Court upheld the High Court's not guilty verdict for all on-duty personnel from CGTD Corporation and LCY Chemical, affirming that they had adequately exercised their duties of care based on the available information during the emergency response. Court judgments found that there were no regulations or established standards for testing pipeline integrity in Taiwan; pipeline test protocols were decided and agreed upon by onsite personnel based on their prior operating experience. This opinion suggested that the practitioners' inadequate knowledge and care for the connecting underground pipelines—invisible yet always interacting with their daily work and engineering systems—were deemed excusable.

Court records further indicate that the negligence of the city government engineers was not an isolated incident but part of a wider pattern of systemic issues. Decades of practice of culvert construction only paid attention to ensuring the function of culverts, rather than the pipelines that interfaced with the underground structures. City-wide inspections after the explosions found that there were more than 4500 pipelines passing through culverts, including 1100 metal pipelines that could degrade because of humid environments. Professionals familiar with culvert construction testified that since culverts are designed to manage water flow, inspectors focus on checking for blockages at the bottom rather than pipelines that may cross at the top. Also, judges noted that the government's engineers were not obligated to initiate or ensure the relocation of the pipeline after they completed the culvert construction in 1992, because such a duty of care was outside the scope of their affiliated office, which dealt with hydraulic engineering and water resources, not pipelines.

4 From Systematic Ignorance to Institutional Reform

4.1 Historical and Political Roots of Collective Ignorance

Although the disaster may be attributed to a series of mishaps and incompetence on the incident day as well as the negligence in culvert construction from decades ago, our above account demonstrated that limited public knowledge about the entanglement of petrochemical pipelines with citizen life had a broad, lasting effect on every practitioner working in the city. In fact, many citizens were similarly unaware of the underground industrial pipelines prior to the explosions, revealing a widespread lack of understanding of the infrastructure and its associated risk among both technical practitioners and the public. A lack of understanding also diminished the capacity of firefighters and chemical emergency response teams to take effective measures for a petrochemical leak.

We argue that the ignorance was a product of the sociopolitical history of Taiwan's technological development policy. The propene pipeline leading to the explosions was planned by the state-owned CPC Corporation as early as 1986, prior to the lifting of martial law. Before CPC Corporation helped Taiwan Polypropylene (succeeded by LCY Chemical in 2008) build the pipeline in the early 1990s, it had long been a founding investor of Taiwan Polypropylene since the latter was established in 1973 and supplied it with propene from CPC Corporation's naphtha cracking plant. During the mid- and late-1970s, the KMT government pushed for the Ten Major Construction Projects, one of which aimed to establish a petrochemical industry in Kaohsiung as a response to the 1973 oil crisis. This was accomplished by expanding CPC Corporation's exclusive business for fractional distillation and cracking, converting imported crude oil into various raw materials (one of them being propene) to supply downstream petrochemical manufacturers. It is noteworthy that Taiwan Polypropylene and its propene pipeline were not unique in its deep connection with the state. State-related, KMT-based investments flowed into nearly half of similar downstream petrochemical manufacturers, influencing their development and business decisions through the control and distribution of state resources and subsidies (Fang et al., 2019; Wang, 1995). The policy was expected to help reduce Taiwan's dependency on petrochemical imports, upgrade the chemical industry, and enable the export of high-value materials.

Taiwan's state-led, top-down approach to the petrochemical industry achieved international success, partly because developed countries, facing ever growing environmental concerns and pollution control costs, had to halt further developments and relocate the pollutive industry to less developed countries with looser environmental and public safety regulations (Wang, 1995). Before the explosions, Kaohsiung had over 1200 kilometers of underground industrial petrochemical pipelines, linking CPC Corporation refineries nearby harbors to various petrochemical industrial parks established in different inland areas. Most of these pipelines were built between 1970 and the early 1990s. No regulations mandated their maintenance or inspection because they transported raw petrochemical materials, and only pipelines of petroleum were regulated.

For most of the period, censorship under military rule discouraged civil society from investigating or discussing the risk of the petrochemical industry, leaving the hazards of underground industrial pipelines invisible and unattended. Even though there were grassroot demonstrations against industrial pollution beginning in the early 1980s, instead of challenging government agencies who were responsible for industrial policy and regulatory inaction, nearly all these environmental movements were direct conflicts with local factory owners by sabotage or sit-ins (Ho, 2001). Generally, the public believed that economic and industrial development was inevitable and desirable, and its externalities were considered a necessary evil. Most individuals did not have knowledge about the risk associated with industrial development. It was only after becoming victims of factory pollution that they recognized their rights to "an inviolable [healthy] body" and "freedom from pollution" (Lii & Lin, 2000).

We found that news media rarely covered the planning or construction of underground industrial pipelines, let alone examining their risk. Instead, most news articles either celebrated milestones of petrochemical production or reported on natural disasters and shortage of raw materials leading to "incidents" of production line shutdown.[2] Even after democratization, demonstrations and public discourse against the petrochemical industry had only addressed the most "visible" issues—industrial pollution from factories and explosion incidents at the facilities. It is unsurprising that the technical practitioners liable for the explosions in the city's ground were unfamiliar with how these off-site, underground pipelines were entangled with their daily practices. A lack of understanding and awareness of hidden pipelines and their risk compromised the ability and readiness of both public engineers and private practitioners to ethically adhere to industry standards and guidelines when making decisions. Ironically, the former authoritarian government's control of public information and knowledge backfired. Its legacy left the democratized government and its agencies incompetent in the emergency response, assuming most of the civil liabilities for the disaster.

4.2 *Reconstructing Governance Oversight and Public Knowledge*

As we saw in our account of the 2014 Kaohsiung propene explosions, the decision space of all the critical actors (including engineers and non-engineers) was constrained by the systemic ignorance of the risk of

[2] Searching through the online database of United Daily News (聯合報)—one of the largest newspapers in Taiwan—we found that there were only five news articles between 1974 and 1992 explicitly discussing or mentioning long-distance petrochemical pipelines.

petrochemical pipelines under the unique sociohistorical contexts of Taiwan, causing all the participants' negligence and inaction prior to the explosions. Their capacity and autonomy were limited in one way or another so that no one could decisively be held accountable. As no engineers could make real decisions, critics may contend that a disaster is not a good case for engineering ethics (e.g., Davis, 2007). Engineers are presumed to "often not have time to wait for a social or even organization-wide solution to the problem that arose" and to "have to act as an individual in the next few hours—or minutes" (Davis, 2006, p. 230). Favoring quick, decisive actions that may admirably save the day, the motivation for such criticism is perhaps attributable to the fact that critics are often not interested in and are perhaps somewhat impatient about the slower-going systemic, institutional change for better governance, regulations, guidelines, etc., that may support ethical decision making.

As a response, we argue that in the context of a democratized society like Taiwan, where social, political, and cognitive infrastructure for enabling and empowering ethical decision making has been inadequate, there are good reasons to prioritize improving governance, policy, public knowledge, and other social institutions rather than being fixated on ascribing responsibilities to individuals. In fact, accountable governance and post-disaster reform after the propene explosions were direly needed to remedy past harm from the entrenched industrial development policy and to support ethical behaviors and decisions of different private and public parties.

As the explosions and their aftermath burst open the black-boxed, hidden infrastructure of petrochemical pipelines and exposed their risk, the city and central government began improving governance and policy. While the judicial system collected facts, determined responsibilities, and perhaps offered a sense of closure to victims and their families, the legislative and executive branches strived to improve the governance of petrochemical pipelines. By the end of 2014, the Industrial Development Bureau of Taiwan developed common guidelines for managing the safety of underground industrial pipelines, covering such topics as emergency response training and pipeline integrity testing that were relevant to the explosions. Locally, the Kaohsiung City Council passed two ordinances in 2015 governing "existing industrial pipelines" (既有工業管線). That same year, the city also established the Office of Pipeline Safety, which always had staff on duty to monitor pipeline safety and respond to emergencies.

These government measures, in turn, raised safety awareness among technical practitioners from all parties, making it more likely that stringent procedures would be swiftly and accurately implemented during an emergency. For example, during an incident of a presumed ethene gas leak in 2020, all the four pipeline operators in the leakage area stopped operations within 30 minutes after being notified by the Office of Pipeline Safety. Within 90 minutes, pipeline operators were directed by authorities to start venting the pipelines by recapturing, flaring, and purging the remaining liquefied gas—a more expensive but safer approach. Throughout the emergency response, pipeline operators and government agencies were responsive and collaborative. By the time pipeline leakage was conclusively ruled out, pipeline owners had undergone continuous pressure tests for over 250 hours.

Importantly, stringent regulatory measures and effective (albeit costly) emergency responses by the government were not merely the result of better, more proactive governance. They were also a product of a functional civil society beginning to take stock of the pipelines' sociopolitical history, such as how government-subsidized petrochemical manufacturers have externalized costs to the environment and risks to citizens, and to engage in citizen oversight informed by a more comprehensive understanding of the industrial infrastructure. While environmental NGOs protested the inaction of the government to transform and phase out petrochemical industrial parks in Kaohsiung, they also recognized the reality and participated in the policy formation of pipeline governance. For example, after the explosions, Citizen of the Earth, Taiwan (CET), an environmental NGO, urged the city government to disclose underground pipeline information and involve citizens in emergency response drills. They also prompted legislative bodies to amend regulations for supervision and accountability of pipeline maintenance. Meanwhile, citizen scholars from the Taiwanese STS community participated in the production of a series of documentaries, *Open the S Files: When Science Meets Social Controversies*, with an episode dedicated to the Kaohsiung Propene Explosions. The series won Taiwan's Golden Bell Award for Best Natural Science Documentary Program in 2019 for informing the public about complex social and policy issues of science and technological development.

5 Conclusion

Ethical practice for public safety does not operate in a vacuum where practitioners follow standards and procedures to the letter. It requires an awareness and understanding of the broader context of practice, including various sociotechnical processes and systems that interact with one's daily work, to make informed decisions that prioritize the community's wellbeing. Despite the presence of petrochemical emergency response procedures and culvert construction standards that could have prevented the disaster, our discussion of the 2014 Kaohsiung Propene Explosions revealed that these measures were collectively compromised and ignored at the practitioners' discretion.

Judges determined "neglect of duty" by establishing whether practitioners "chose" not to follow more stringent existing procedures and "acted" out of expediency and convenience. Nevertheless, reading between the lines of court judgments, we found that courts acknowledged that there were no better choices available, because practitioners had limited knowledge—no more than that of average citizens—about how long-distance industrial pipelines interacted with their daily work and engineered systems. This systematic neglect and ignorance of the infrastructure's inherent risks was the leading contributing factor to the disaster, as reflected in court findings. By contrast, the 2020 ethene gas leak prompted a rapid, coordinated emergency response, as pipeline operators and government agencies recognized the significant risk to the urban community and took immediate, effective actions.

Our analysis revealed that micro-ethical issues encountered by individual engineers were simultaneously macro-ethical challenges of governance and technological development—a position in engineering ethics that is distinctly native in Taiwan (Shih, 2021). Consequently, education and professional development aiming to improve ethical practice should prepare citizen practitioners with an understanding of the sociopolitical history of technological development that shapes engineering practices and systems. In a democratized society like Taiwan, it is also important to recognize that efforts to foster a sociopolitical understanding of technology among practitioners for supporting their ethical practice, in turn, derive their strength from a political landscape and public sphere that are empowered, disposed, and welcoming to make the sociopolitical processes of technology visible and accountable. This includes a commitment to transformative justice that acknowledges past harm, ensures accountable governance, discusses and incorporates lessons learned, and sustains post-disaster reconstruction and reform. To improve public awareness, open government and data transparency were crucial first steps highlighting the entanglement of the petrochemical industry and the city's dense population. For instance, at the end of 2015, the city government published a mapping system including route information for industrial and natural gas pipelines. The system allowed the public to track a pipeline's route from one factory to another on a map, showing the streets and intersections it traversed as well as the pipeline's owner, the petrochemical substances being transported, and its decommissioning status.

Ten years after the explosions, the Kaohsiung City mayor publicly recognized that Kaohsiung's past cooperation with national economic development policies had passed the risks of industrial pipelines on to its citizens—albeit a concern that had long been repeatedly expressed by civil society. While the government planned to relocate the scattered petrochemical industrial parks to southern, coastal Kaohsiung, environmental NGOs and activist scholars advocated for the transformation and phasing out of the petrochemical industry. These civilian efforts to raise public awareness would make it less likely for practitioners to detach from the extensive sociotechnical processes and systems that are both the context and foundation for duty of care. Where citizens engage with the petrochemical industry, grassroots political movements and social forces serve as catalysts for the advancement of proactive ethical practice, much as they do in driving the formulation and application of laws and regulations (Adams et al., 2023).

Lastly, we should note that although our discussion of the 2014 Kaohsiung Propene Explosions is rooted in the sociopolitical context of democratized Taiwan and its petrochemical sector, the lessons we discussed would extend to other cases of ethical practice, including in technoscience and beyond. Since ethics-related standards, guidelines, and procedures are driven by a profession's expanding connections with a diverse network of sociotechnical systems and stakeholders, commitment to ethics in practice needs to be supported by knowledge and awareness about social factors and contexts that are entangled in these relationships, which may initially seem "nontechnical" or "external" to the profession. Historical and political aspects of professional practice may even become more salient when a former lack of public understanding or discourse has a lasting impact that hinders the development or support of ethical behaviors. For example, beyond our case in Taiwan, ethical practices that are related to principles of justice, equity, diversity, and inclusion would never gain enough strength from rules or compliance. They require civil society and government to inform the public about past harms and the importance of remedies, thereby making ethics more likely to become an integral part of a citizen practitioner's daily practice.

References

Adams, J., Schütz, T., & Fortun, K. (2023). Late industrialism, advocacy, and law: Relays toward just transition. *East Asian Science, Technology and Society: An International Journal, 17*(4), 462–493. https://doi.org/10.1080/18752160.2023.2267338

Anonymous. (1986, April 24). 工程倫理與環境保護 [Engineering ethics and environmental protection]. *United Daily News*, p. 2.

Chang, K.-H., Downey, G. L., & Shih, B. P.-J. (2025). *Engineers and the two Taiwans: The abnormal club*. Springer.

Chang, L.-S. 張隆盛. (1987, January 1). 公共建設應重視社會成本 [Public construction should take societal costs seriously]. *United Daily News*, p. 14.

Chao, L., & Myers, R. H. (1998). *The first Chinese democracy: Political life in the Republic of China on Taiwan*. Johns Hopkins University Press.

Chen, H.-H. (2023). Field report: Taiwan's RCA litigation and its multiple outreaches: The experience of an STS community, 2011–2023. *East Asian Science, Technology and Society: An International Journal, 17*(4), 494–520. https://doi.org/10.1080/18752160.2023.2269019

Chin, W.-S. 金文森, & Chiang, C.-H. 江政憲. (2017). *工程倫理 [Engineering ethics]*. Wunan.

Chou, C.-H. 周卓煇. (2014). *工程倫理: 以工程, 醫學, 商業等個案解讀 [Engineering ethics: An introduction through case studies in engineering, medicine, and business]*. Gau Lih.

Chu, P.-Y. (2010). 徘徊在《新史學》邊緣:臺灣的科技史研究 [Wandering on the fringe of *new history*: Research into the history of science and technology in Taiwan]. *Newsletter for Research in Chinese Studies, 29*(3), 1–7.

Cutcliffe, S. H. (2000). *Ideas, machines, and values: An introduction to science, technology, and society studies*. Rowman & Littlefield.

Davis, M. (2006). Engineering ethics, individuals, and organizations. *Science and Engineering Ethics, 12*(2), 223–231. https://doi.org/10.1007/s11948-006-0022-y

Davis, M. (2007). Perils of using Hurricane Katrina to teach engineering ethics. *IEEE Technology and Society Magazine, 26*(4), 16–22. https://doi.org/10.1109/MTS.2007.911074

Fang, H.-C. 房慧真, Ho, J.-S. 何榮幸, Lin, Y.-I. 林雨佑, & Chiang, I.-T. 蔣宜婷. (2019). *煙囪之島:我們與石化共存的兩萬個日子 [A smoking island:*

Petrochemical industry, our dangerous companion more than fifty years]. SpringHill Publishing.

Herkert, J. R. (2005). Product safety and liability. In C. Mitcham (Ed.), *Encyclopedia of science, technology, and ethics* (pp. 1508–1512). Macmillan Reference USA.

Ho, M.-S. 何明修. (2001). 台灣環境運動的開端:專家學者, 黨外, 草根 (1980–1986) [The inception of environmental movements in Taiwan: Scholar-experts, political dissidents, and grassroots organizers (1980–1986)]. *Taiwanese Sociology*, *2*, 97–162. https://doi.org/10.6676/TS.2001.2.97

Hong, W.-L. 洪文玲, & Wang, J.-P. 王治平. (2022). 蟄伏的惡龍?城市地下石化管線的治理 [A dragon lurking below? Governing urban underground petrochemical pipelines]. In W.-Y. Lin, W.-H. Kuo, H.-Y. Wang, & K.-Y. Yang (Eds.), *科技社會人4:跟著關鍵物去旅行 [Traveling with STS things]* (pp. 30–49). National Yang Ming Chiao Tung University Press.

Hong, W.-L. 洪文玲, Wang, J.-P. 王治平, & Lu, N.-X. 盧念萱. (2018). 站在氣爆的十字路口:談多種工程專業的交, 錯 [At the crossroads of explosions: The intersection and conflict of multiple engineering works]. In W.-Y. Lin, T.-T. Lin, & K.-Y. Yang (Eds.), *科技社會人3:跨領域新驛路 [Technology, society, and humans: A new path toward interdisciplinarity]* (pp. 169–180). National Chiao Tung University Press.

Hsiao, H.-H. 蕭新煌. (1981, March 21). 「疏離感」與「第六倫」 ["Sense of alienation" and "the sixth relationship"]. *United Daily News*, p. 2.

Li, K.-T. 李國鼎. (1981, March 28). 經濟發展與倫理建設 [Economic development and ethics formation]. *United Daily News*, p. 2.

Lii, D.-T., & Lin, W.-Y. (2000). 社會力的文化根源:論環境權感受在台灣的歷史形成 (1970–86) [The cultural origins of social forces: The historical formation of perception of environmental rights in Taiwan (1970–86)]. *Taiwan: A Radical Quarterly in Social Studies*, (38), 133–206. https://doi.org/10.29816/TARQSS.200006.0004

Lynch, W. T., & Kline, R. (2000). Engineering practice and engineering ethics. *Science, Technology, & Human Values*, *25*(2), 195–225. https://doi.org/10.1177/016224390002500203

Martin, M. W., & Schinzinger, R. (1983). *Ethics in engineering*. McGraw-Hill.

National Science Council. (1991). *第四次全國科學技術會議結論 [The conclusion of the 4th national science and technology conference]*. National Science Council.

Natsume, K. (2021). *Japan's engineering ethics and Western culture: Social status, democracy, and economic globalization*. Lexington Books.

Pan, K.-C. 潘國正. (1992, December 5). 交大開設工程倫理課程因應複雜政治環境及權力結構 [Chiao Tung University offers engineering ethics courses as a response to the complex political environment and power structure]. *China Times*, p. 15.

Petroski, H. (1985). *To engineer is human: The role of failure in successful design*. St. Martin's Press.

Rigger, S. (1999). *Politics in Taiwan: Voting for democracy*. Routledge.

Shih, B. P.-J. (2021). Towards an engineering ethics with non-engineers: How western engineering ethics may learn from Taiwan. In Z. Pirtle, D. Tomblin, & G. Madhavan (Eds.), *Engineering and philosophy: Reimagining technology and social progress* (pp. 161–180). Springer Nature. https://doi.org/10.1007/978-3-030-70099-7_8

Shih, P.-J. B., & James, M. (2024). Thinking through engineering: How our values about engineering shape engineering ethics education. *Teaching Ethics, 24*(1), 145–163. https://doi.org/10.5840/tej202442143

Tai, D.-Y. 戴東源, & Hong, W.-L. 洪文玲. (2018). 台灣工程倫理回顧兼論工程論理的科技與社會取徑 [A review of engineering ethics in Taiwan from the STS perspective]. *Taiwanese Journal for Studies of Science, Technology and Medicine*, (26), 185–232. https://doi.org/10.6464/TJSSTM.201804_(26).0004

Teng, J.-T. 鄧治東. (1991). 中原大學開設工程倫理課程的經驗與評估 [The experience and evaluation of the engineering ethics course offered by Chung Yuan Christian University]. *Science Development Monthly, 19*(3), 319–321.

Wang, H.-S. 王晃三. (1991). 談工程倫理教育 [On engineering ethics education]. *Science Development Monthly, 19*(3), 316–318.

Wang, H.-S. 王晃三. (1992). *大學「工程倫理」課程的教學設計之研究 [A study on instructional design for college engineering ethics courses].* National Science Council.

Wang, H.-S. 王晃三. (1994). *溶入各工程專業課程的倫理教學設計 [Design for ethics instruction that permeates through each professional engineering course].* National Science Council.

Wang, J.-H. 王振寰. (1995). 國家機器與臺灣石化業的發展 [The state and the development of Taiwan's petrochemical industry]. *Taiwan: A Radical Quarterly in Social Studies,* (18), 1–37. https://doi.org/10.29816/TARQSS.199502.0001

Wei, C. 魏誠. (1981, March 24). 建立新倫理 [Developing new ethics]. *United Daily News*, p. 5.

Weil, V. (1984). The rise of engineering ethics. *Technology in Society, 6*(4), 341–345. https://doi.org/10.1016/0160-791X(84)90028-9

Wilcox, J. R. (1983). The teaching of engineering ethics. *Chemical Engineering Progress, 79*(5), 15–20.

Yang, H.-N., Chen, J.-H., Chiu, H.-J., Kao, T.-J., Tsai, H.-Y., & Chen, J.-R. (2016). Confined vapor explosion in Kaohsiung City—A detailed analysis of the tragedy in the harbor city. *Journal of Loss Prevention in the Process Industries, 41*, 107–120. https://doi.org/10.1016/j.jlp.2016.03.017

Yeh, W.-K. 葉文冠, Chen, T.-H. 陳德興, & Huang, T.-Y. 黃鼎元. (2017). *科技與工程倫理 [Technology and engineering ethics].* Wunan.

CHAPTER 5

Technology and Microscopy in Biomedical Education: Navigating the Ethical Landscape

R. Claire Aland, Rebecca Lush, and Nicole Shepherd

1 Introduction

This chapter outlines our approach to addressing ethical questions in our day-to-day work as educators using collections of human tissue. A recent repurposing of teaching spaces has prompted us to engage with several collections of human tissue used for teaching. Our collections include material of unknown provenance with no documented consent, raising questions about whether this material should be used, and if so, by whom, and how. Even where consent for use has been provided, developments in

R. C. Aland (✉) • R. Lush
School of Biomedical Sciences, The University of Queensland, Brisbane, QLD, Australia
e-mail: r.aland@uq.edu.au; r.lush@uq.edu.au

N. Shepherd
Rural Clinical School, University of Tasmania, Burnie, Tas, Australia
e-mail: nicole.shepherd@utas.edu.au

J. Hayes, S. Maslen (eds.), *Ethics in the doing of technoscience*,
https://doi.org/10.1007/978-3-032-20002-0_5

technology allow a range of uses not possible when materials were collected. Digital materials can be easily distributed, whereas when originally collected, physical specimens would not have been seen or used outside of a physical teaching space. Additionally, to offer quality biomedical education in a dynamic educational, scientific, clinical, and technological landscape, materials must be generated to keep collections current. Sound ethical management is necessary for maintaining public trust in the institution, which is vital if we wish people to donate material. It also allows the institution to meet its mission and obligations. This chapter outlines existing regulatory, legal, and ethical frameworks. We did not find adequate guidance that addressed both legal and ethical aspects of day-to-day collection management. Therefore, we propose an ethical framework for managing collections of human tissue, particularly in the intersection with technology. We then offer examples from our work to demonstrate how this framework can be applied by staff working with similar collections.

1.1 Background

Globally, many institutions hold collections of materials derived from or manufactured by humans. These include objects humans have manufactured (tools, art) or human bodies (skeletons, tissues, organs, whole bodies). This chapter focuses on material originating from human bodies that we term ***human tissues*** or ***material of human origin***.[1] Substantial risks already exist for institutions using materials of human origin in teaching and research. These include reputation loss through public perception of unethical conduct (Rodgers, 2003), or legal and financial repercussions through knowingly or inadvertently performing illegal activities (Zinn, 2007). As paths of inquiry develop in education and research, there may be requests to use existing material in new ways or to collect new material for new uses. Digital technology allows instant and unregulated worldwide dissemination of images derived from materials of human origin.

[1] We use this term to center the humanity of the person from whom the material was collected, similar to the term ***medical products of human origin*** (Martin, 2017). Other terms in use are ***human biological materials*** (e.g., Meslin & Quaid, 2004; Martin, 2011). We use the term origin or originator rather than donor to refer to this person, to highlight that consent for collection may not have been obtained.

These new uses expose institutions to pre-existing and new risks (Comer, 2022).

In our roles, we have been considering the ethical implications of our pedagogical practice and praxis.[2] We are currently transforming both practice and praxis toward these collections by considering stewardship as outlined by Golde (2006) and Cleas and Deblon (2017). Historically, human tissue within similar collections has been anonymized, becoming objects *owned* by an institution or individual collector, exercising the rights that come with possession, such as use and disposal (Falconer, 2019). Today, however, a broader emphasis on collective custodianship emphasizes that multiple stakeholders (researchers, curators, donors, and communities) must take responsibility for this material (Parry & Gere, 2006). We do not use the words *caretaker* or *custodian* as these terms imply ownership and/or maintenance of existing collections. Golde's (2006) conceptualization of disciplinary stewardship includes not only responsibilities for conservation and transformation of existing materials but also the generation of new materials.

We will illustrate our implementation of this approach, with three case studies: virtual microscopy technology, use of artificial intelligence (AI) technology, and genetic and genomic technologies. These case studies demonstrate how our ethical framework is interwoven in our praxis as educators, now, and in preparing for a transformative future.

Practices of collecting material of human origin are informed by legal frameworks and social norms, which change over time. Change can challenge the purpose and use of institutional collections, as often these collections have been amassed over decades or centuries. For some collections, provenance might be unknown, as information on materials' origins was either not recorded or was lost in the intervening years (Cornwall et al., 2024). Even when provenance is known, it can be unclear whether the person from whom the material originated had legal capacity and gave their full consent (using our contemporary understanding of informed consent) for collection, retention, and use of their tissues (Comer, 2022). For example, for material obtained prior to the Declaration of Helsinki in 1964, it is unlikely that a documented consent process was followed. There is

[2] Using Mahon et al. (2020), we define practice as actions that we (educators and students) perform routinely and habitually in relation to the material while praxis encompasses actions based on consideration of the consequences of performing them and driven by a desire to contribute in a positive and meaningful way to society (something greater than ourselves).

precedent of waiving consent for the use of human biological material in research where it is difficult or impossible to contact the donors (Gefenas et al., 2011), and this could be extended to using such materials in education.

This lack of documented consent introduces an ethical strain when using materials in biomedical education, where students are otherwise taught that informed consent is fundamental to scientific research and clinical practice. It requires us to navigate an uncomfortable path of "do as we say, not as we do". Downplaying the (lack of) consent for teaching materials suggests the requirement for consent might be flexible, with potential adverse outcomes for our students' future interactions with other humans. We argue that human tissues have special ethical significance to originators, their families, and communities. If we choose to use these materials for education, informed consent for use and retention, or a lack of such, should be explicitly centralized in our ethical teaching praxis (Cornwall et al., 2024).

The University of Queensland, Australia, where we work, has multiple, geographically separate, campuses. Historically, separation of collections has and continues to introduce complexity to their management and use. The pathology collection is held in the Integrated Pathology Learning Centre (IPLC) at the University's Herston campus. The museum is open to the general public and is used by students studying biomedical science and medicine (The University of Queensland, 2005) (see Fig. 5.1). The collection has approximately 5000 organs (in preserving fluid in Perspex pots), plastinates (treated with plastic resin), and 57,000 histopathology slides (small pieces of tissue prepared for microscopy). The histology collection is housed at the University's St Lucia campus, 8kms from Herston. This slide collection contains many tens of thousands of slides of normal and near-normal human and animal tissue. The histology collection has never been accessible to the general public.

Our collections were created for education, their raison d'être to promote and satisfy curiosity, and enable the experience of wonder and awe. Initially, that experience was limited by the time and space in which students could interact physically with the collections. Now, technology gives us transformative ways to see and use our collections, and drives the generation of new materials. While initially physical, our collections are now also inextricably digital (see Fig. 5.2). Technology helps us to create innovative, digitally based learning experiences for our current students, facilitating awe, wonder, and exploration of curiosity, which is essential to science education (Hadzigeorgiou, 2011, 2016; Lindholm, 2018). Digital images of slides are now globally available online on a not-for-profit,

Fig. 5.1 The Integrated Pathology Learning Centre, The University of Queensland, Herston campus

Fig. 5.2 Technology examples at the School of Biomedical Sciences, The University of Queensland, St Lucia Campus

education-focused image bank and virtual microscope [https://www.best.edu.au]. While not physical, images are derived from and represent people, raising questions about access, relationships between physical and digital versions of specimens, storage and disposition of physical specimens once digitized, and disposition of the digital versions (IFAA, 2023).

In the next section we will outline the current ethical and legal situation for our specific context. While broad guidelines are available, there is a need to develop a new ethical framework to guide the use of our collections for teaching and research, particularly for newly emerging uses that center technology. We will outline this framework and then apply it to three case studies from our institution, highlighting how it can support decision making for existing and new applications of technology to materials of human origin.

2 Existing Regulatory Frameworks and Gaps

2.1 Legal Frameworks

In general, laws provide a baseline legal framework that has sufficient scope for people to derive ethical standards appropriate to their context rather than acting as proscriptive authority. Body donation, where people donate their whole bodies with informed consent, following their natural deaths, for teaching and research in anatomy, occurs in many countries around the world. However, materials of human origin in our pathology and histology collections do not come via this route but rather from tissues removed during hospital autopsies or from living individuals during surgery or other medical procedures. In Australia, the Royal College of Pathologists of Australasia issued guidelines of the use of human tissue (2022), which state that when tissue is removed with consent for legitimate research or medical need (such as surgery or medical procedures, as our collections), the individual of origin no longer has rights to that tissue as part of their own body. Tissue can additionally be removed during an autopsy and kept for testing (Queensland Government, 2003). These guidelines are similar to those used internationally (Petrini, 2012). Under the Queensland *Coroners Act 2003*, tissues of such origins are considered to be property of the institution holding them, but the legislation is often too general to provide useful guidance on specific operational issues (Queensland Government, 2003).

At the time of writing, the primary legal framework applying to our collections is the *Transplantation and Anatomy Act 1979* (Qld)

(Queensland Government, 1979). The Act is divided into ten parts covering Schools of Anatomy, donations of tissue after death and prohibiting a trade in tissue. The Act uses the terms *respect* and *dignified treatment* without accompanying contextual definitions. Neither pathology nor histology collections fall under the legal scope of the School of Anatomy license which governs the Body Donor Program at our institution, and are therefore covered by only a portion of the Act.

In addition to this Act, there are strong legal frameworks surrounding tissue that is, or most likely originates, from a First Nations (Aboriginal or Torres Strait Islander) person, which must, in Queensland, be reported to the Department of Women, Aboriginal and Torres Strait Islander Partnerships and Multiculturalism for investigation (Queensland Government, 2003).

However, management needs to be more than just what is legally permissible. While these frameworks outline the fundamental requirements for legally managing a collection containing human tissues in our jurisdiction, they do not adequately address contemporary ethical standards and norms. A range of standards and guidelines have been developed by organizations that provide oversight and guidance that go some way to filling this gap.

2.2 Collection Management Frameworks

The IPLC is a museum and uses existing museum frameworks to guide operations and management. The histology collection and IPLC additionally draw on various ethical frameworks created for or applied to human tissue in other contexts. Collection management frameworks follow themes of institution, provenance, and community. These three themes not only scaffold the AAA Recommendations for the Management of Legacy Anatomical Collections (Cornwall et al., 2024) but are also raised in the two standards documents that outline best practice and considerations for museum management in Australia. These are the *National Standards for Australian Museums and Galleries 2.0* (National Standards Taskforce, 2023) and *International Council of Museums (ICOM) Code of Ethics for Museums* (International Council of Museums Ethics Committee, 2017). These frameworks all state they are to be implemented while also adhering to context-specific local and national laws. Technology is included in many documents but generally as a means of record-keeping for collections, rather than as a primary means for access, generation, and transformation.

A common theme throughout all frameworks is the necessity for institutional governance structures, policies, and procedures. The benchmark for governance best practice and collection management includes, but is not limited to, collection management systems, auditing, and necessary documentation (National Standards Taskforce, 2023; International Council of Museums Ethics Committee, 2017). For collections containing material of human origin, the first section of the ICOM Code of Ethics covers how to respectfully store and collect "in a manner consistent with professional standards and the interests and beliefs of members of the community … from which the objects originated, where these are known" (International Council of Museums Ethics Committee, 2017, p. 11). The AAA recommendations (Cornwall et al., 2024) build on this with their operational domains, focusing on the physicality of this material and how it should be documented. Similar to the other two frameworks, this refers to inventory processes and record-keeping, and the use of technology in such. All three frameworks directly link institutional governance with the next theme of provenance as a critical component of collection management.

Provenance is raised more generally in the AAA recommendations when compared with commonly adopted museum frameworks that typically reference this issue only in relation to First Peoples Ancestral Remains. In the *National Standards for Australian Museums and Galleries 2.0* (National Standards Taskforce, 2023), First Peoples Ancestral Remains are the only materials of human origin covered in the policy with the legalities and repatriation processes outlined. For example, in Section C there is a standard that "the organization develops an understanding of, and implements, its responsibilities to, the care, handling, storage and/or return of First Peoples Ancestral Remains and Secret/Sacred Cultural material held in collections" (National Standards Taskforce, 2023, p. 14). The AAA recommendations are broader in their approach to provenance, encouraging collections to consider using genetic information to establish provenance to potential ancestral communities.

The final theme, community, covers providing collective custodianship of material, its storage, use, and public display. This situates material within a contemporary community of care, which includes cultural and historical communities, ethical interests and/or genetic relatedness (Cornwall et al., 2024, p. 2789), which may or may not overlap with the community of origin. The concept of community is also covered in the ICOM *Code of Ethics for Natural History Museums* (International Council of Museums

International Committee for Museums and Collections of Natural History, 2013), which states that material must be displayed with dignity and collaboration with origin communities should be sought. As with the other themes, the AAA provides a more specific framework for this theme, making an important distinction between stakeholders (those with a specific interest who may not identify as part of a community of care, such as students) and communities of care (those interested in the management of the material) (Cornwall et al., 2024, p. 2789).

Overall, these frameworks provide both general procedural and ethical guidance but are not comprehensive guides. In addition, all three documents use terms such as dignity and respect without clear definitions or providing a framework to support their application in collection management. None of them consider technology past its role in management of physical materials.

In our context, the IPLC has developed local policies and ethical guidelines for managing the collection by combining museum frameworks, elements of the AAA recommendations and relevant legislation. The histology slides were not aggregated into a collection until 2022, so processes are less developed; nevertheless, we are following the IPLC's leadership. This recent aggregation of the histology slides has been a driver for the development of our ethical framework.

2.3 Summarizing Regulatory Gaps and the Need for a New Framework

Examining existing regulatory frameworks demonstrates gaps in the literature surrounding material of human origin. Notably, there is no framework for material from living or non-living humans that was not donated through an existing body donor program requiring pre-mortem informed consent for use and retention from donors. These frameworks do not provide sufficient guidance for evaluating existing or potential uses. The intersection of technology with our collections, as a means of generation and transformation, rather than simply record-keeping, has enabled uses inconceivable to collectors, originators, or donors. Specifically, digitization creates digital forms of physical specimens which are not like-for-like. Guidelines designed for larger and discrete, individual specimens (Cornwall et al., 2024) do not account for the unique distributed and unlimited nature of physical microscopic tissue or their digital versions. Before outlining our framework, we want to highlight aspects of our collections that make it challenging to apply existing guidelines.

Firstly, our microscopic collection contains human tissue that is often small and not immediately recognizable as human. There is no rule or guideline in Australia for how much of a person is considered human. Almost all would agree that an intact deceased body could be considered human and should be treated with respect, but that for most cultures, hair cut at the hairdresser would not demand the same respect. A useful framework should demonstrate, contextually, why all material of human origin should be treated consistently.

Secondly, the provenance for much of our collections is untraceable. Some have attached records, but for the remainder it is impossible currently to identify originators. We can surmise, from when, where, and how it was collected, that most of this material is white and not representative of today's diverse Australia. We acknowledge significant historical contributors to the non-representative and exclusionary way in which collections such as ours were often formed and which continue to contribute to inequality in our teaching and in healthcare that people receive today. A useful ethical framework for guiding continuing and future use of our collections must acknowledge this and seek to redress past lack of representation and exclusion (Jenkins, 2010). This is particularly important for building trust in communities that have suffered exploitation and dispossession (Moodley & Singh, 2016), particularly in scientific research and healthcare. Our collections currently exist within a multi-cultural and spiritually diverse society, different from when they were originally collected. A useful ethical framework must then be able to recognize and include contrasting and diverse cultural and spiritual views in managing the collection, rather than imposing the viewpoints of one group within society on all donors.

Thirdly, our collections were deliberately obtained for teaching and research, rather than incidentally aggregated through collector's curiosity (Alberti, 2011). The purpose of our collections therefore differs from many other collections of materials of human origin. However, as already highlighted, there is the issue of consent or lack thereof. A useful ethical framework must explicitly center expectations around informed consent and the educational uses of the material.

Related to this is final disposition. Today, there might be pressure to dispose of human tissue where there is a lack of documented consent (Cleas & Deblon, 2017). Disposal may be proposed to avoid adverse publicity (Wright & Hibbitts, 2024). If we consider this material to be human, then disposing simply to avoid a problem is an unethical manifestation of "out of sight, out of mind". Removing physical material also removes their educational and dialogical possibilities. Use can encourage personal reflection and consideration of the cultural contexts that shape the history of education and scientific research. Digital technology use raises further issues. While controlling a single large physical specimen for disposal or withdrawal is straightforward, multiple small slides and their digital versions are far more difficult to track and, if necessary, remove from circulation. This highlights the need for proactive ethical consideration of possible outcomes of actions rather than reactionary damage control.

For teaching and research collections like ours, there are additional ethical considerations of how needs would be met without access to the material. Not even new forms of technology can replace the collections (yet). Technology, however, provides more equitable access to existing collections, so technology even simply as a means of access raises ethical questions in relation to students and communities that they will later serve as healthcare practitioners.

Our framework will function as a bridge between these general frameworks and our specific circumstances. By doing so, our framework can holistically guide current practice, accommodate future practices, and be applied in other contexts. This ethical framework is consistent with local law and flexible enough to incorporate varied cultural considerations and accommodate future technological developments.

3 Developing an Ethical Framework for Collection Management

To address gaps in existing regulatory frameworks, we looked for guidance in existing ethical frameworks. The best-known set of principles used in biomedicine is the four principles approach developed by Beauchamp and Childress (2019), which outlines four clusters of principles: respect for autonomy, non-maleficence, beneficence, and justice. These principles were proposed in order to provide guidance in bioethical dilemmas. While this framework has been widely applied to a variety of ethical problems in biomedicine, it does not provide an adequate framework for managing our collections.

We found that consideration of the principles of reverence and dignity provided additional guidance beyond this four principles approach (Beauchamp & Childress, 2019). Reverence was included in doctors' training in the early twentieth century (Schweitzer, 1923; Mostwin, 2019) but has fallen out of favor. Woodruff (2014) argues for the importance of reverence as a secular concept and traces it across cultures and times in history. He explains that reverence is "an ancient virtue that survives among us in half-forgotten patterns of civility, in moments of inarticulate awe, and in nostalgia for the lost ways of traditional cultures" (Woodruff, 2014, p. 1) and defines it as "the well-developed capacity to have the feelings of awe, respect, and shame when these are the right feelings to have" (Woodruff, 2014, p. 6). He outlines three aspects to reverence: a sense that there is something larger than an individual human being; accompanied by capacities for awe, respect, and shame; and often expressed in, and reinforced by, ceremony. It does not belong to religion, but it is an important virtue that strengthens ties between individuals and their community. The use of material of human origin in education or public displays is a form of connection between living and deceased persons; therefore, using this material is an occasion for reverence (Jones, 2019).

Similarly, dignity is an important concept that builds on the intuition that all human beings are entitled to respect and to have the ability to shape their lives in the way they wish (Nussbaum, 2011). Although dignity is often linked to the concept of personhood (i.e., debates about the point in human development when a collection of cells and tissues is considered a person), Kirchhoffer and Dierickx (2011) conclude that tissue by itself cannot be said to have dignity, but it is implicated by these materials and therefore how we approach this tissue is worthy of moral consideration. Similarly, Martin (2011) argues that human biological materials possess moral value by virtue of their connection to persons. We consider that dignity is relevant to collections of material of human origin, considering the personhood of originators/donors of tissue, the connection to the community, and humanity more broadly. It is ethically justifiable to consider this as human and treat this consistently with how we treat other humans (Cleas & Deblon, 2017). Example 5.1 describes one such case.

Example 5.1

In 2016 a polycystic liver was donated to the IPLC by a living donor. The donor wanted to share her story to raise awareness of the impact of organ donation. The IPLC now has the liver on display alongside a presentation that shares her story and the message of organ donation.

Conceptions of dignity are culturally bound. Our institution anonymizes anatomy donors, as most in Western countries do (Jones & King, 2017; Greene & Rosen, 2019). However, non-Western approaches to anatomy education, such as the Silent Mentor Program that originated in Taiwan, include donors' identifying information such as their name, life story, family, and community (Lin et al., 2009; Winkelmann & Güldner, 2004; Douglas-Jones, 2017). In the Western tradition, anonymity is presented as respect for donors but also as protection for students (Sukol, 1995), removing subjectivity and emotional involvement in favor of objectivity and emotional detachment. The body is viewed as a specimen rather than person and respect is centered around the choice to donate (while alive) rather than any intrinsic human qualities of the body itself. The body instead has an extrinsic or instrumental value as a teaching tool. Both knowledge (biomedical science) and humanistic qualities are necessary for students of biomedicine as we describe in Example 5.2.

Example 5.2

We designed an undergraduate teaching module that uses histology and IPLC collections in both physical and digital forms. Students learn normal structure and function by considering the impact of pathology (dysfunction) on people's lived experiences, taking neuroanatomy from the theoretical learning of individual students to a broader narrative including the lives of people in the surrounding community.

The emphasis on (objective) knowledge over (subjective) humanistic qualities in education (Pedersen, 2010) is demonstrated by students' decreasing empathy levels as they progress through training (Rosenfield & Jones, 2004). In contrast, approaches like the Silent Mentor Program center the donor's personhood through including stories about their life, involving the donor's loved ones and community in students' education. In this way, respect and gratitude for donors is demonstrated not only through acknowledgment of their choice to donate but also through honoring their life story, their family, and their community. The emphasis on respect, gratitude, and recognition of intrinsic value (dignity) has informed our approach to developing an ethical framework for our collections.

From Western philosophy, we can see similar ideas in Nussbaum's conception of dignity as intertwined with respect, agency, and equality—all beings (including non-human animals) should be treated with respect and enabled to use their full range of capabilities (Nussbaum, 2006, Claassen, 2014). A key capability is affiliation—to live with and show care for others. Decisions to donate material to these collections for the education of tomorrow's scientists and healthcare professionals are important expressions of affiliation with the community (Farsides et al., 2023). Similarly, Kirchhoffer and Dierickx (2011) suggest that persons donate bodily material to be morally good and thereby derive a sense of self-worth from donation. This altruistic desire of a donor to contribute can be considered when making decisions about materials of human origin. Respecting donors' intentions demonstrates respect for human dignity. Our recognition of dignity inherent in the tissue in our collections centers it as part of a human community, rather than solely through its value as an anonymized, dehumanized resource (Douglas-Jones, 2017).

Example 5.3 provides a specific scenario illustrating our approach incorporating all framework principles. In this case, we created guidelines for student use of images and photography of both collections.

Example 5.3

Students bring their personal devices into the IPLC and use them in the vicinity of the physical material to produce images. They can also digitally access our virtual microscope and image bank from any location with their login credentials. A challenge we faced was providing guidance to students on image generation and use outside supervised learning environments. To resolve student confusion over circumstances in which images could be used and shared, we created guidelines for taking photographs and screenshots. These guidelines highlight the inherent dignity of the material by reminding students that it is of human origin and explicitly stating what constitutes respectful behavior. The guidelines remind students of beneficence (using images in their learning rather than for entertainment) and non-maleficence (sharing the images within a teaching and learning context to avoid distressing others).

These examples illustrate how education and public display can demonstrate respect to the originators of the material, promote wonder at the complexity of the human body, and also connect with our shared human experiences: health, vulnerability to disease, and ultimately, death.

4 Our Ethical Framework in Practice

4.1 *Ethical Practice with New Technologies*

With rapid development of new technologies, including artificial intelligence, other uses for the collections have been proposed, which raise new ethical issues. Following our assumption of inherent dignity for this material arising from its humanity, we apply the four biomedical ethical principles approach (Beauchamp & Childress, 2019) in a prima facie manner to further critique our practice and inform our future praxis.

The materials in our collection were collected and are held legally, but without a requirement for documented consent, many slides do not have information on provenance or informed consent and thus do not meet current best practice in anatomy teaching or research. We cannot say that respect for autonomy has been upheld (the rights of an individual to exert control over their body) in creating our collections, raising the question of whether we should continue to hold, and use, these collections. We can consider the remaining prima facie principles, beneficence, non-maleficence, and justice for the originators of the tissue, the stakeholders (students, staff, institution), and the wider community. By adding the principle of reverence, our focus can turn to considering how the community could be involved in making decisions about the use and management of collections, for example, by including religious rituals appropriate for the originators. The following sections offer three examples to illustrate our approach.

4.1.1 *Current Use of Virtual Microscopy Technology*

Over the past two decades, virtual microscopy has been implemented worldwide, including our institution, for histology and histopathology teaching, research and clinical practice. Digitization results in digital images existing alongside physical slides. At the time it was introduced there was little consideration of ethical questions or potential outcomes.

Users and the wider community derive substantial benefit from the use of our collections in education and their subsequent contribution to clinical care and research. Due to the unique biological nature of these collections, it is currently impossible to derive this benefit otherwise. The use of virtual microscopy technology contributes an additional significant benefit by supporting more equitable access (justice) than permitted by the physical collections. Virtual microscopy technology allows students and staff to benefit from using a technique used in research and clinical practice worldwide. Another benefit of virtual microscopy technology is that digital images reduce institutional burdens of storage and maintenance of physical specimens and microscopes.

However, there may be harm caused to cultural groups from retention, display, and use of tissue from their members. If we continue using material collected without informed consent or require different behavior from students or staff depending on whether human material is collected with informed consent or not, we undermine and harm the importance of obtaining consent in research and healthcare. Because these collections are not representative of the modern Australian population and are also not inclusive for biological sex and age, portraying them as representative risks creating and reinforcing unconscious bias in students and staff (Queensland Places, 2001). This may contribute to poorer health outcomes for under-represented groups and the possibility of bias in research.

Virtual microscopy, through the unlimited number and distribution of digital copies, amplifies these potential harms more than using only physical material. If concerns about any particular images are raised, it is near impossible to control or recall, meaning that harm may continue indefinitely. Additionally, information associated with images such as species, disease, sex, and consent for use may be lost in distribution, leading to images being used out-of-context, which may produce additional harm. Harm has great potential to generate community distrust, impacting our ability to generate new material that is more representative and inclusive. At an institutional level, we are mitigating harm through recognition and involvement. Our university formally recognizes the need for collection caretakers through creating positions of Specialist Collection Officers. This role is assigned caretaking and stewardship of histology and histopathology collections, including storage and auditing. We also mitigate harm through formal training for our students and community outreach. The institution has recognized that histology and pathology collections should be managed in alignment with other collections at the University, through the establishment of an oversight committee. In terms of community outreach, we are establishing trust with our wider community, welcoming people into the space and demonstrating transparency with our collection.

While the existence and continued use of the collections with virtual microscopy technology results in benefits that would be impossible to obtain otherwise, there are significant risks of harm and perpetuating injustice, which need addressing and mitigation.

4.1.2 Future Use of Artificial Intelligence Technology

Artificial intelligence (AI) is increasingly used in research and clinical practice using histology and histopathology. Deep learning AI analyzes and extracts data from images and performs visual object recognition. In histology, this means analyzing digital slides for the presence of normality or pathology, which requires large training sets of normal and pathology slides (Dippel et al., 2024). The large numbers of slides in our collections can function as these large training sets, potentially benefiting future research and clinical practice.

Incorporating AI within our teaching will benefit students by providing them with relevant skills and knowledge, and also the community through better training of future clinical professionals or researchers. This proposed use also highlights the importance of ethical curation of these collections to realize these future benefits.

Potentials for harm are similar to virtual microscopy (given the similarity of technology), but this does not preclude additional harm. The ability of AI to quickly process and extract data relating to tissue from a particular individual, or group of people, may exaggerate existing biases within the collections. This can result in greater harm to under-represented groups. A training set composed purely of white samples may result in AI not recognizing pathology in samples originating outside of that subset (false negatives) or conversely may detect pathology when there is none (false positives). Both of these outcomes may harm people who are not represented in the training set; their disease may progress further without detection or be subject to additional unnecessary investigations to rule out false positives. As these may affect people who historically, and today, are under-represented or have poorer outcomes when interacting with the healthcare system, this use of technology may further perpetuate inequity (injustice).

4.1.3 Future Use of Genetic and Genomic Technologies

Some tissue in our collections could be processed to extract genetic information. Genetic testing may yield information about diseases that could benefit individuals or communities at the national or global levels.

The extraction of genetic information, coupled with increasing use of online genetic libraries, may allow re-identification of material that has hitherto been anonymous, in a process similar to solving cold cases, where collection samples could be linked back to genetically related individuals who have deposited their DNA information into these libraries. If samples are re-identified, it may conceivably benefit the originator of the tissue, their immediate family, or those genetically related to them, if particular genetic information might inform their decisions about healthcare.

However, there may also be the risk of harm if originators or their loved ones do not know that parts of their body are used in these collections and become distressed by that knowledge. Additionally, they may have chosen not to investigate genetic elements to diseases in themselves or their families, and being given that information may affect their lives detrimentally, such as being denied health insurance.

Overall, application of our ethical framework supports the retention and use of the collections are beneficial, although harm or injustice may arise through the use of non-consented material in our teaching, or because the collections do not represent the modern Australian population and may perpetuate historical inequity in healthcare and outcomes. The current uses of technologies with the existing collections are beneficial and equitable (just), although harm or injustice is possible because of a lack of control over the distribution of digitized material. Proposed uses of new technologies with existing collections may benefit, but also cause harm or injustice, through extraction of biological information, the scope of which is not currently clear, and which may impact the originators or donors of the tissue. Identifying potential sources of harm and injustice allows us to avoid, mitigate, and repair past harm, which upholds the inherent dignity of the material.

4.2 Dignity and Reverence in Practice

The inclusion of dignity and reverence as foundational ethical principles in collection management and education invites consideration of spiritual, cultural, and societal aspects to the use of human tissue. Broadening the use of our collections to include humanities alongside biomedical science education and research helps to situate the content and practice of science within its social, cultural, and historical context. This includes exploring why the collections are non-representative, and how historical bias may continue as unconscious bias and impact the quality of people's lives and their interactions with healthcare providers.

As stated above, our framework sets a foundation for respect and gratitude for materials when the originator had no part in, and probably, no knowledge of the uses of their body parts once they were separated from them. Where materials have documented consent from donors, gratitude and respect is demonstrated by acknowledging the generosity of the donor and their loved ones. The donor has been altruistic with their body (respect for autonomy), and gratitude is shown for the gift to students and teachers to facilitate their learning (beneficence). Students and staff are expected to display respectful behavior toward the material to avoid disrespecting the donor's actions (admiring the donor for their beneficent qualities) or harming the donor's loved ones (by treating their family member with dignity and acknowledging the gift through public ritual). In contrast, the materials in our pathology and histology collections where there is no identified donor and/or no documented consent, the disconnect between the living originators, their loved ones, the community they belong to, their culture, and what happens in the classroom is even greater. By including dignity and reverence for the originators of the material, a wider foundation of respect can be built that goes beyond simply a respect for autonomy and donor's wishes.

Educators can assist the development of respect arising from a genuine sense of admiration, rather than being performative to meet expectations and avoid punishment. Framing respectful behavior around inherent dignity allows respect to emerge from a regard for and admiration of our shared humanity with the originators of the materials. Students and staff experience and create reverence through their interactions with those materials. Woodruff (2014) described how staff and students can co-create the reverent classroom through mutually respectful behavior. Our "reverent classroom" is created in the intersections of students, staff, human tissues, the communities surrounding those collections, and our shared experiences of humanity, regardless of where, when, or what form that classroom takes (physical or digital). Gratitude emerges from reverence, from recognition of our shared humanity and the educational benefits accessed through the material. Staff and students share in an experience of awe and wonder as they make the connection with common humanity and life larger than themselves as individuals. An important demonstration of reverence is recognition through ceremony. In our collections, this is expressed in the manner in which materials are cared for, used, and ultimately disposed. These ceremonial behaviors allow us to express reverence and reinforce its centrality to the stewardship of our collections.

Curiosity and wonder are central to learning and to scientific discovery. However, curiosity and wonder by themselves are insufficient to guide ethical behavior. Curiosity and wonder drove the formation of many collections of materials of human origin around the world, including ours. When materials are considered as property, there is likely to be a lack of respect for the originators of the material, or their cultures or beliefs, or recognition of the inherent dignity of humanity in the material. Woodruff's inclusion of shame as part of reverence is very apposite when viewers are faced with historical collections of human material. These were likely collected in ways that do not meet contemporary community standards or perpetuated discrimination against marginalized groups of people. It is important to be transparent about this history in order to acknowledge past wrongs. Shame then, as well as awe and wonder, becomes necessary to the experience of reverence with materials of human origin in our reverent classroom.

5 Conclusion

Our framework allows us to work from an imperfect starting point—where scientists with curiosity and drive to understand the human body created collections of human material, but these are lacking documentation of provenance and consent. More recently, materials were digitized without consideration of potential harms. Rather than hide this, as educators it is important to encourage reflection and engagement as part of understanding the history of science and the cultural contexts of the production of scientific knowledge. Curiosity and wonder have created many collections of materials of human origin over the centuries, but this very human desire to seek knowledge and satisfy curiosity must be tempered with ethical conduct.

In this chapter, we outlined how existing legal and ethical frameworks did not provide adequate guidance for our practice in collection stewardship. We present a new framework and give examples of how it guides our practice and praxis. Recently, we established an institutional oversight committee to oversee our work. We continue using our collections to meet our educational and research obligations but now openly acknowledge the lack of consent in our educational practice. Our expectations of use now center around the principles of respect, dignity, and reverence.

For the future, we aim to create trust through ethical management allowing us to collect material generating a more representative collection which meets future needs. We can then withdraw non-consented material from use and establish a means for reverent disposition that acknowledges the contribution to education and research. This will allow us to continue meeting the original purpose of our collections, to create experiences of awe and wonder for future generations of students and researchers, while building on an ethical foundation.

References

Alberti, S. J. M. M. (2011). *Morbid curiosities.* Oxford University Press.

Beauchamp, T. L., & Childress, J. F. (2019). *Principles of biomedical ethics* (8th ed.). Oxford University Press.

Claassen, R. (2014). Human dignity in the capability approach. In M. Duwell, J. Braarvig, R. Brownsword, & D. Mieth (Eds.), *The Cambridge handbook of human dignity* (pp. 240–249). Cambridge University Press.

Cleas, T., & Deblon, V. (2017). When nothing remains: Anatomical collections, the ethics of stewardship and the meanings of absence. *Journal of the History of Collections, 30*(2), 351–362.

Comer, A. R. (2022). The evolving ethics of anatomy: Dissecting an unethical past in order to prepare for a future of ethical anatomical practice. *The Anatomical Record, 305*(4), 818–826. https://doi.org/10.1002/ar.24868

Cornwall, J., Champney, T. H., de la Cova, C., Hall, D., Hildebrandt, S., Mussell, J. C., Winkelmann, A., & DeLeon, V. B. (2024). American Association for Anatomy recommendations for the management of legacy anatomical collections. *The Anatomical Record, 307*(8), 2787–2815. https://doi.org/10.1002/ar.25410

Dippel, J., Prenissl, N., Hense, J., Liznerski, P., Winterhoff, T., Schallenberg, S., Kloft, M., Buchstab, O., Horst, D., Alber, M., Ruff, L., Müller, K.-R., & Klauschen, F. (2024). AI-based anomaly detection for clinical-grade histopathological diagnostics. *NEJM AI, 1*(11), AIoa2400468. https://doi.org/10.1056/AIoa2400468

Douglas-Jones, R. (2017). 'Silent mentors': Donation, education, and bodies in Taiwan. *Medicine Anthropology Theory, 4*(4), 69–98.

Falconer, K. (2019). Dismantling 'doodeward': Guided discretion as the superior basis for property rights in human biological material. *University of New South Wales Law Journal, 42*(3), 899–927.

Farsides, T., Smith, C. F., & Sparks, P. (2023). Beyond "altruism motivates body donation". *Death Studies, 47*(1), 56–64. https://doi.org/10.1080/07481187.2021.2006827

Gefenas, E., Vilius, D., Cekanauskaite, A., & Serepkaite, J. (2011). Research on human biological materials: What consent is needed, and when. In C. Lenk, J. Sandor, & B. Gordijn (Eds.), *Biobanks and tissue research* (pp. 95–110). Springer.

Golde, C. M. (2006). Preparing stewards of the discipline. In C. M. Golde & G. E. Walker (Eds.), *Envisioning the future of doctoral education: Preparing stewards of the discipline* (pp. 3–20). Jossey-Bass.

Greene, S. J., & Rosen, L. (2019). Sharing personal information about anatomical body donors: What first-year medical students want to know and how it affects emotional responses to dissection. *Clinical Anatomy 32*(8), 1019–1032. https://doi.org/10.1002/ca.23389

Hadzigeorgiou, Y. (2011). Fostering a sense of wonder in the science classroom. *Research in Science Education, 42*(6), 985–1005. https://doi.org/10.1007/s11165-011-9225-6

Hadzigeorgiou, Y. (2016). *Imaginative science education: The central role of imagination in science education.* Springer.

International Council of Museums Ethics Committee. (2017). *ICOM code of ethics for museums.* https://icom.museum/wp-content/uploads/2018/07/ICOM-code-En-web.pdf

International Council of Museums International Committee for Museums and Collections of Natural History. (2013). *ICOM code of ethics for natural history museums.* https://icom.museum/wp-content/uploads/2018/07/ICOM-code-En-web.pdf

International Federation of Associations of Anatomists. (2023). *Recommendations for good practice around human tissue image acquisition and use in anatomy education and research.* Retrieved September 16, 2025, from https://ifaa.net/committees/ethics-and-medical-humanities-ficem/recommendations-for-good-practice-around-human-tissue-image-acquisition-and-use-in-anatomy-education-and-research/

Jenkins, T. (2010). *Contesting human remains in museum collections: The crisis of cultural authority.* Routledge.

Jones, D. G. (2019). The dead body as an object of investigation, intrigue, and reverence. In J. Fitzgerald & A. Moyse (Eds.), *Treating the body in medicine and religion* (pp. 197–214). Routledge.

Jones, D. G., & King, M. R. (2017). Maintaining the anonymity of cadavers in medical education: Historic relic or educational and ethical necessity? *Anatomical Science Education, 10*(1), 87–97. https://doi.org/10.1002/ase.1618

Kirchhoffer, D. G., & Dierickx, K. (2011). Human dignity and human tissue: A meaningful ethical relationship? *Journal of Medical Ethics, 37*(9), 552–556. https://doi.org/10.1136/jme.2010.041509

Lin, S. C., Hsu, J., & Fan, V. Y. (2009). "Silent virtuous teachers": Anatomical dissection in Taiwan. *BMJ, 339*, b5001. https://doi.org/10.1136/bmj.b5001

Lindholm, M. (2018). Promoting curiosity? *Science & Education, 27*(9), 987–1002. https://doi.org/10.1007/s11191-018-0015-7

Mahon, K., Heikkinen, H. L., Huttunen, R., Boyle, T., & Sjølie, E. (2020). What is educational praxis? In K. Mahon, C. Edwards-Groves, S. Francisco, M. Kaukko, S. Kemmis, & K. Petrie (Eds.), *Pedagogy, education, and praxis in critical times* (pp. 1–20). Springer. https://doi.org/10.1007/978-981-15-6926-5_2

Martin, D. E. (2011). *Beyond the market: A new approach to the ethical procurement of human biological materials.* Doctoral dissertation, The University of Melbourne.

Martin, D. E. (2017). Ethical aspects of medical products of human origin. *ISBT Science Series, 12*(1), 281–287. https://doi.org/10.1111/voxs.12304

Meslin, E., & Quaid, K. (2004). Ethical issues in the collection, storage, and research use of human biological materials. *Journal of Laboratory Clinical Medicine, 144*(5), 229–234. https://doi.org/10.1016/j.lab.2004.08.003

Moodley, K., & Singh, S. (2016). "It's all about trust": Reflections of researchers on the complexity and controversy surrounding biobanking in South Africa. *BMC Medical Ethics, 17*(57). https://doi.org/10.1186/s12910-016-0140-2

Mostwin, J. L. (2019). Reverence for the body: An ethical principle grounded in human experience. In J. Fitzgerald & A. Moyse (Eds.), *Treating the body in medicine and religion* (pp. 87–98). Routledge.

National Standards Taskforce. (2023). *National standards for Australian museums and galleries 2.0.* https://magsq.com.au/wp-content/uploads/2021/02/NSFAMG-SECOND-EDITION-2023.pdf

Nussbaum, M. C. (2006). *Frontiers of justice: Disability, nationality, species membership.* Belknap Press.

Nussbaum, M. C. (2011). *Creating capabilities: The human development approach.* Belknap Press.

Parry, B., & Gere, C. (2006). Contested bodies: Property models and the commodification of human biological artefacts. *Science as Culture, 15*(2), 139–158. https://doi.org/10.1080/09505430600708036

Pedersen, R. (2010). Empathy development in medical education - A critical review. *Medical Teacher, 32*(7), 593–600. https://doi.org/10.3109/01421590903544702

Petrini, C. (2012). Ethical and legal considerations regarding the ownership and commercial use of human biological materials and their derivatives. *Journal of Blood Medicine, 3*, 87–96. https://doi.org/10.2147/JBM.S36134

Queensland Government. (1979). *Transplantation and Anatomy Act 1979* (Qld). https://www.legislation.qld.gov.au/view/html/inforce/current/act-1979-074

Queensland Government. (2003). *Coroners Act 2003* (Qld). https://www.legislation.qld.gov.au/view/html/inforce/current/act-2003-013

Queensland Places. (2001). *Herston.* https://queenslandplaces.com.au/herston

Rodgers, L. (2003). Human bodies, inhuman uses: Public reactions and legislative responses to the scandals of bodysnatching. *Nottingham Law Journal, 12*(2), 1–17.

Rosenfield, P. J., & Jones, L. (2004). Striking a balance: Training medical students to provide empathetic care. *Medical Education, 38*(9), 927–933. https://doi.org/10.1111/j.1365-2929.2004.01931.x

Royal College of Pathologists Australasia. (2022). *Ethical and legal issues in relation to the use of human tissue in Australia and New Zealand.* https://www.rcpa.edu.au/getattachment/ad0d4af1-901a-4056-af05-56bbc06167c1/Ethical-and-Legal-Issues-in-Relation-to-the-Use-of.aspx#:~:text=The%20traditional%20notion%20is%20that,therapeutic%20reason%2C%20testing%20or%20donation

Schweitzer, A. (1923). *Civilization and ethics* (Vol. 2). A. & C. Black.

Sukol, R. B. (1995). Building on tradition of ethical consideration of the dead. *Human Pathology, 26*(7), 700–705. https://doi.org/10.1016/0046-8177(95)90216-3

The University of Queensland. (2005). Integrated pathology learning centre: School of biomedical sciences. https://biomedical-sciences.uq.edu.au/facilities/integrated-pathology-learning-centre

Winkelmann, A., & Güldner, F. H. (2004). Cadavers as teachers: The dissecting room experience in Thailand. *BMJ, 329*(7480), 1455–1457. https://doi.org/10.1136/bmj.329.7480.1455

Woodruff, P. (2014). *Reverence: Renewing a forgotten virtue.* Oxford University Press.

Wright, J. R., Jr., & Hibbitts, W. A. (2024). Dick van Velzen and the Burnside warehouse organ scandal in Nova Scotia. *Canada. Clinical Anatomy, 37*(1), 102–113. https://doi.org/10.1002/ca.24115

Zinn, C. (2007). Sydney medical dean apologises to relatives for misuse of body parts. *BMJ, 334*(7589), 335. https://doi.org/10.1136/bmj.39125.460440.94

CHAPTER 6

The Limits of Ethical Engineering Actions: The Titan Case

Jan Hayes, Rune Storesund, and Sarah Maslen

1 Introduction

High reliability research reminds us that in complex, hazardous systems, safety and reliability depend on organizations that defer to technical expertise, rather than hierarchy, for high-stakes, time-pressured technical decisions (Weick & Sutcliffe, 2001). When an aircraft in flight has a safety issue, the pilot has authority to act without reference to airline management on the ground. Many organizations that operate hazardous technology have similar processes in place for operations (Klein, 1999), but, in contrast, when it comes to major engineering-driven decisions such as design choices involving major resource commitments, formal authority typically sits at the top of the organization chart, not with those closest to

J. Hayes (✉) • S. Maslen
School of Property, Construction and Project Management, RMIT University, Melbourne, VIC, Australia
e-mail: jan.hayes2@rmit.edu.au; sarah.maslen@rmit.edu.au

R. Storesund
Center for Catastrophic Risk Management, University of California, Berkeley, CA, USA
e-mail: rune@storesundconsulting.com

J. Hayes, S. Maslen (eds.), *Ethics in the doing of technoscience*,
https://doi.org/10.1007/978-3-032-20002-0_6

the technology. The accident record shows that important safety messages are often lost in transmission through multiple organizational actors with varying priorities, an issue that is countered through organizational structures that link technical experts directly to decision makers (Hopkins, 2019).

Other reasons for communication failure between technical experts and decision makers may be cultural, such as when bad news reaches decision makers but is not welcomed (Hayes & Maslen, 2022, 2023). Nevertheless, when engineers sense that something is wrong, when data or intuition warn that risk is rising, the responsibility to speak up comes not just from company policy or legal duty but from the core of what it means to be a professional (Hayes et al., 2021). Acting on that responsibility, however, is rarely simple. Past accident cases such as the Deepwater Horizon blowout (Reader & O'Connor, 2014), the Challenger and Columbia space shuttle losses (CAIB, 2003; Rogers, 1986; Starbuck & Farjoun, 2005), and the Boeing 737 MAX crashes (Committee on Transportation and Infrastructure, 2020; Hopkins, 2025) show that speaking up for safety in a professional setting can mean challenging authority, confronting indifference, or risking professional isolation. It requires moral courage, patience, and skill in communication—qualities that are easy to admire in hindsight but can be difficult to summon in the moment.

While many organizations seeking a positive safety culture encourage speaking up for safety, in practice concerns raised by technical professionals are not always well received (Paul et al., 2025; Whitmore & Keeler, 2025). When professionals find concerns are ignored, they sometimes choose to move to other courses of action including leaving the organization. Indeed, high turnover can indicate a serious disconnection between technical workers and senior management on safety issues (Hayes et al., 2024). Another possible course of action is whistleblowing, i.e., reporting safety issues externally, typically to a regulatory body, when standard internal processes for raising concerns have failed (Schmidt, 2020). External whistleblowing such as this is "the last line of defense" offering an intervention in the face of fraudulent, illegal, and even willfully negligent practices when other systems and controls have failed (Motarjemi, 2023, p. 1091).

We seek to explore the tipping point between one set of behaviors focused on collaboration and organizational learning, and other behaviors aimed at disconnecting an engineer from organizational lines of authority. We ask: what drives an engineer with safety concerns to move from being an insider working on solutions to an organizational outsider? What are the implications for managing major accident risk?

The story of the 2023 Titan submersible implosion illustrates just how high the stakes can be for organizations and for individual experts. Titan's crew of five died instantly when the carbon fiber hull catastrophically imploded in the deep ocean on an expedition to visit the wreck of the Titanic. In the wake of this tragedy, the US Coast Guard (USCG) initiated a formal Marine Board of Investigation (MBI). This study draws on evidence presented as part of the MBI[1] and the MBI's final report (Whalen & Neubauer, 2025). Our review of this material identified three technical professionals who had significant safety concerns and tried to convince OceanGate's CEO to act on their advice. Their warnings were met not with dialog but with denial and, in some cases, retaliation. Our analysis of their experience including their departure from OceanGate underscores an uncomfortable truth: taking the high road—doing what is right rather than what is easy—sometimes means walking away.

We argue that for professional engineers, having an awareness of the complex dynamics of speaking up is crucial. Speaking up is not an act of disloyalty; it is a core expression of professionalism and public trust. The challenge, in part, lies in doing so effectively: framing the message clearly, grounding it in evidence, and connecting it to shared organizational goals. Nevertheless, where organizational goals such as cost or schedule consistently trump value-driven concerns about system safety, the best option for experts may be to withdraw from the organization. This choice is strongly influenced by social embeddedness within the workplace as well as professional values. Our review of the Titan case shows a complex relationship between these factors which is worthy of further research attention.

2 Professional Values in Technoscientific Work

Engineering does not occur in isolation. Every engineer works as part of a social and organizational system that conceives, builds, and sustains technology. Work with technology is typically a collective, social activity with many hands involved (van de Poel et al., 2012), where contributions to design, operation, and maintenance are distributed across many professionals, each contributing specialized knowledge and judgment. While

[1] See https://www.news.uscg.mil/News-by-Region/Headquarters/Titan-Submersible/

technical competence is indispensable, the social context in which engineers work both enables and constrains how that competence is applied (Reason, 1997; Vaughan, 1996). As such, engineering work is never value-neutral: choices regularly reflect trade-offs among safety, efficiency, and cost (Aven, 2009; Jones-Lee & Aven, 2011). Exercising professional judgment means integrating technical facts with human values, and this is precisely where engineering ethics intersects with safety and risk management (Maslen et al., 2021). By understanding their place in this wider sociotechnical network, engineers can more effectively translate technical insight into informed, ethically grounded contributions that help steer organizations toward safer outcomes (Johnson, 2020). A review of the literature highlights three practical ways individual engineers can use their expertise, communication skills, and professional judgment to help their organizations become safer, more reliable, and more adaptive.

The first practical behavior is paying attention to weak signals. Many engineers seek out small faults, failures, and anomalies such as an unexpected reading, a design deviation, or a recurring maintenance issue for their value as early indicators of larger system weaknesses (Hayes & Maslen, 2023). Bringing these observations forward promptly and seeking early action, even if they seem minor, is key in preventing small technical problems from becoming organizational crises.

Sometimes weak signals are an indication of a more significant problem that requires an appeal to decision makers that control the necessary resources to resolve. In cases like this, effective communication is the key to initiating action. When an engineer becomes aware of a high-stakes issue, the way in which advice is framed is critical in turning a potential conflict into a collaborative improvement effort. Expert engineers possess a deep and detailed understanding of how system components interact, where vulnerabilities lie, and how small technical choices can cascade into major safety outcomes (Hayes & Maslen, 2023). This safety imagination, i.e., the ability to link actions to the potential for adverse outcomes in the future (Pidgeon & O'Leary, 2000), means that engineers can be not only problem solvers within a technical domain but also ethical participants in a shared enterprise whose decisions have social consequences. Reminding decision makers that choices can have major ethical consequences is therefore a key professional behavior in speaking up for safety (Hayes, 2015). Conversely, while the link to consequences is important, focusing only on this can be viewed as an attitude that "the sky is falling". Explaining persuasively to decision makers why they should care must also include presenting data grounded in organizational goals such as reliability, compliance, or efficiency and being solution-oriented (Hayes et al., 2025).

Finally, the literature on engineering liability advises engineers to document reasoning and recommendations in detail (Hayes et al., 2017). In addition to providing some measure of liability protection, a clearly argued analysis including supporting evidence and clear recommendations to decision makers increases the likelihood that action is taken. Written communication creates organizational memory and accountability—two cornerstones of a learning culture (Reason, 1997).

These behaviors are all intended to bring about organizational action to fix issues of concern, but this is not always achievable. Engineers understand that they are not passive actors in the system and have a responsibility to act, but that there are limits (Hayes et al., 2025). Organizational power relationships are typically used as the explanation as to why employees choose not to speak up or to back down if concerns are not acted upon immediately (Hopkins, 2019). Sometimes obstacles to organizational learning may be so significant that whistleblowing may be required to address "corporate myopia" regarding safety issues (Pidgeon & O'Leary, 2000, p. 24). By stepping outside organizational lines of authority and responsibility, whistleblowing is viewed as a threat to dominant cultural norms. In some cases whistleblowers have been subject to severe reprisals from their employers, the state and even society more generally (Munro et al., 2025). In the context of safety, Stene and Eikemo (2020) found that whistleblowers can be undermined publicly by their colleagues, that it can negatively affect their work assignments, and that they may experience overt bullying and even disciplinary action. It is little wonder then that in many cases professionals choose to leave and find work in an organization that is culturally more aligned with their values, rather than continuing to engage across a large cultural divide (Gonzalez & Pérez-Floriano, 2015; Maurer & Qureshi, 2019).

There is a substantial literature on "safety voice", but engineers are a special case of the broader question of organizational actors speaking up for safety. As a recent review shows, the research focus has been on front line workers speaking up about issues that pose an immediate safety threat (Paul et al., 2025). Safety voice is described as "proactive communication aimed at preventing physical harm" (Paul et al., 2025, p. 1), which does not preclude engineers raising concerns about imagined future safety problems, but speaking up is not widely studied from this perspective.

Overall, we see that the safety literature describes various behaviors that are likely to be effective for engineers in speaking up for safety. Some aim to foster collaboration and maximize influence within the boundaries of organizational authority. Other behaviors require a shift to moving outside such boundaries and withdrawing from the organization. What is missing is a safety literature that addresses the transition one to the other. The Titan case provides an opportunity to begin to address this gap.

3 The Titan Submersible

3.1 *OceanGate's Origins*

OceanGate's founding vision in 2009 was bold and ambitious: to revolutionize human access to the deep ocean through innovation and affordability. The company sought to build a small fleet of deep-diving submersibles capable of carrying five people to depths of up to 6000 meters and to make them available for charter anywhere in the world. By designing submersibles that could be transported in standard shipping containers and launched from ships of opportunity at local ports, OceanGate aimed to dramatically reduce costs, increase flexibility, and make deep-ocean exploration accessible to a broader segment of humanity. In doing so, OceanGate positioned itself as a disruptor, challenging long-standing norms in undersea engineering and operations in pursuit of a grand vision that fused adventure, science, and accessibility.

3.2 *Developing Titan*

As OceanGate evolved, the company made a deliberate shift from conventional industry-accepted vessels toward a far more ambitious technological goal—developing a new class of deep-diving submersibles that were lighter, easier to transport, and capable of reaching extreme depths. OceanGate's CEO, Richard Stockton Rush, was the driving force behind the company who envisioned a vehicle that would not only dive deeper than existing crewed submersibles but also redefine the economics of deep-ocean access. This vision culminated in the development of Titan, a submersible featuring a carbon fiber hull intended to carry crew to the wreck of the Titanic, nearly 4000 meters below the surface.

In pursuit of this innovation, OceanGate chose to reject long-standing industry conventions that had governed the safe design, testing, and certification of manned submersibles. Beginning in 2012, the company collaborated with the University of Washington's Applied Physics Laboratory (APL-UW) and consulted with Boeing engineers to assess the feasibility of using carbon fiber composites for the hull. Both organizations ultimately withdrew from the partnership, citing fundamental safety concerns associated with OceanGate's departure from established industry-standard engineering protocols. Undeterred by these warnings, OceanGate pressed forward independently, working instead with smaller, more accommodating contractors to fabricate Titan's carbon fiber hull. The company's leadership framed this defiance of conventional practice as intentional disruption—a necessary step to innovate where others had been constrained by caution. OceanGate's approach was thus not an accident of oversight but a conscious decision to challenge the boundaries of accepted marine engineering in pursuit of speed, cost-efficiency, and technological distinction.

3.3 *Titan I (2016–2019)*

OceanGate developed two iterations of the Titan submersible craft. The Titan I hull was manufactured by Spencer Composites. It operated from 2016 to 2019 and was used to initially test deployment of the carbon fiber pressure hull. Dive logs show that Titan I reached a maximum depth of 4000 meters on June 26, 2018, and a manned dive to a depth of 3939 meters on December 10, 2018, with Mr. Rush as the sole crew member. Titan I was retired from service in 2019 due to fracturing of the carbon fiber pressure hull during further test dives in the Bahamas. On this occasion, no-one was hurt.

3.4 *Titan II (2020–2023)*

In 2020, OceanGate sought to strengthen the credibility of its engineering program by contracting with NASA to assist in the development and testing of a new carbon fiber pressure hull. The partnership was publicly framed as a collaboration with one of the world's most trusted technical institutions, lending an air of legitimacy to OceanGate's ongoing work. In practice, however, NASA's involvement was minimal and largely advisory, limited by COVID-19 travel restrictions to remote consultations on a one-third–scale mock-up of the new hull.

Two of these one-third–scale models were manufactured using a single-cure composite layup process. Both failed under pressure testing—one catastrophically, and the second test was aborted just before failure after clear signs of structural instability. Despite these outcomes, OceanGate did not repeat the testing program or systematically analyze the root cause of the failures. Instead, the company proceeded to full-scale fabrication of the Titan II.

Components from the earlier Titan I, including the titanium end domes and interface rings, were reused, and the finished hull was fitted with strain gauges and acoustic emission sensors intended to provide real-time monitoring of stress and microfracture activity. A polyurethane coating was applied to the hull's exterior, which prevented any direct visual inspection of the composite surface. The completed Titan II underwent a single pressure test in March 2021 at the Deep Ocean Test Facility in Maryland, reaching just 1.09 times its intended operating pressure—a safety factor far below accepted standards for manned submersibles.

Following this limited validation, Titan II was mobilized to St. John's, Newfoundland and conducted a series of dives to the wreck of the Titanic in 2021 and 2022. The final Titan II is shown in Fig. 6.1. Out of 23 total dive attempts, only 13 were successful. Each expedition was supported by chartered vessels continuing OceanGate's mission of "disrupting" the traditional model of deep-ocean exploration. Yet beneath the language of innovation lay a consistent pattern: untested materials, reused

Fig. 6.1 The Titan submersible. (Source: US Coast Guard Report of the Marine Board of Investigation into the Implosion of the Submersible Titan.)

components, and reduced safety margins presented as acceptable trade-offs in the pursuit of progress.

3.5 *Safety Warning Signs Unheeded*

OceanGate's tragedy was not the result of an unforeseeable accident but a knowable and preventable failure. From the outset, the company elected to operate outside the framework of established submersible engineering practices that had kept humans safe at great depths for decades. Central to this decision was the choice not to submit Titan—or any of OceanGate's submersibles—to independent certification, a process universally regarded in the industry as essential to verifying a vessel's fitness for operation. Third-party certifying agencies such as the American Bureau of Shipping or DNV-GL typically conduct rigorous, independent design reviews, materials testing, and pressure-chamber validation to confirm structural integrity. Mr. Rush dismissed these protocols as unnecessary bureaucracy—too costly, too slow, and, in his view, an impediment to innovation.

In rejecting certification, OceanGate effectively eliminated an external industry-standard safeguard that would have identified the critical design flaws in Titan's carbon fiber pressure hull. OceanGate possessed test data indicating that the composite material exhibited premature fatigue and inadequate safety margins under cyclic loading, but rather than redesigning or retesting, OceanGate proceeded with operations. The absence of regulatory oversight in international waters left ultimate responsibility in the hands of the company itself—and specifically with its CEO—who chose to press ahead despite repeated warnings from internal engineers, industry experts, and former collaborators.

3.6 *The Implosion*

In June 2023, Titan II imploded, instantly killing its crew, during a deep-ocean dive to the wreck of the Titanic. Titan II imploded approximately 1 hour and 45 minutes into its planned 10 hour dive at a depth of approximately 3500 meters. An extensive international search and rescue operation was launched, but the submersible's wreckage was found approximately 500 meters from the bow of the Titanic. All five occupants onboard were confirmed deceased. Figure 6.2 shows part of the wreckage.

The MBI reported that "OceanGate's failure to follow established engineering protocols for safety, testing, and maintenance of their

Fig. 6.2 Remotely operated vehicle image of TITAN aft tail cone. (Source: US Coast Guard Report of the Marine Board of Investigation into the Implosion of the Submersible Titan.)

submersible, was the primary causal factor" (Whalen & Neubauer, 2025). The investigation further highlighted the poor governance and dangerous workplace culture at OceanGate and the lack of effective regulatory oversight as contributing factors. Regarding Mr. Rush's decision making as CEO, the MBI noted that the "complete absence of any regulatory compliance for the Titan is considered by the MBI to have been a deliberate strategy that progressively prioritized innovation and operational flexibility over passenger safety" (Whalen & Neubauer, 2025, pp. 292–293).

4 Speaking Up for Safety

The testimony of three individuals who worked for OceanGate in the lead up to the Titan implosion gives us an opportunity to investigate how technical professionals act when they have concerns for safety that extend beyond their direct responsibility.

4.1 Director of Marine Operations (2016–2018)

An experienced maritime professional with a deep background in submersible and marine operations was hired as the Director of Marine Operations (DMO) for OceanGate in 2016. His primary responsibilities included overseeing the safety and operational readiness of OceanGate's submersibles, including the Titan. The DMO was deeply involved in conducting test dives, ensuring the operational readiness of critical components, pilot training, system maintenance, producing operational procedures and training manuals, meeting paying clients and liaising with external bodies with an interest in operations such as the support vessel captain and port authorities. He was attracted to work at OceanGate because of the chance to contribute to an exciting, new project that he saw as credible due to the involvement of APL-UW and Boeing.

Once he started with the company, the DMO found that the links to these bodies were not as strong as he had been led to believe. Further, he had been told in interview that he would be the project manager for the design of the Titan I submersible with the design work being under his direct supervision so that he could leverage his operational knowledge and ensure the designed system maximized operational safety. In practice, this was not the way that work at OceanGate was organized. Engineering was largely undertaken by Mr. Rush supported by the Director of Engineering (DE) with little input from the DMO.

Despite his lack of decision making authority, in such a small organization the DMO was aware of design and construction decisions that were being made for Titan I, and he quickly became concerned about several safety issues, particularly the use of carbon fiber for the pressure hull and the lack of comprehensive testing to confirm the hull's structural integrity. His testimony highlighted his attempts to raise these concerns, drawing on his previous submersible experience (more than anyone else in OceanGate). He tried to alert Mr. Rush to the ways in which OceanGate's plans did not comply with industry standards. In line with what he understood to be his professional responsibility, he asked for an explanation as to how the safety of the Titan crew was to be assured. His concerns were dismissed in several ways: by ignoring them, by instructing him not to speak to specialist contractors such as APL-UW, and by face-to-face intimidation.

The safety concerns culminated when he was asked to complete an inspection on the Titan I at the point of the handover of the vessel from

Engineering to Operations. He produced a report in January 2018 detailing 25 specific safety/integrity concerns. His cover note states (in part):

> Verbal communication of the key items I have addressed in my attached document have been dismissed on several occasions, so I feel now I must make this report so there is an official record in place. ... the items noted in my report are significant in nature and must be addressed. As stated in my employment contract ... [I am] responsible for ensuring the safety of all crew and clients during submersible and surface operations. I feel we all have a duty of care as Management to all personnel ... therefore it is my opinion that until suitable corrective actions are in place and closed out, Cyclops 2 (Titan) should not be manned during any of the upcoming trials.

His employment at OceanGate was terminated the following day after a long meeting at which it was clear that the DMO and Mr. Rush were at an impasse; the DMO thought it was unacceptable to put the crew and passengers at risk by diving to unsafe depths in the Titan, whereas Mr. Rush wanted to go ahead with plans to dive at the Titanic site.

The DMO thought about sending a copy of his report to the OceanGate Board saying,

> I didn't do this out of badness. I didn't do this out of any ulterior motive. I did this so that all my voiced concerns were put down on paper and they could go "Wow I see where he's coming with this. There's a lot of things in this report that could affect the integrity of this hull".

Within a week of finishing work at OceanGate in January 2018, he had made initial inquiries regarding making a complaint to the Occupational Safety and Health Administration (OSHA), citing significant safety concerns about the Titan submersible. Subsequently a formal whistleblower complaint was lodged by the DMO to OSHA who also notified the US Coast Guard. The governing legislation (Seaman's Protection Act) meant that this was framed as being about retaliation by OceanGate for the DMO's safety-related actions. While the DMO was concerned that he had lost his job, his primary concern centered on the safety issues, i.e., the structural integrity of the submersible's carbon fiber hull, and the danger it posed to the crew. His primary aim in lodging the complaint was not seeking mediation with OceanGate over his employment but rather triggering the authorities to investigate safety at OceanGate.

OceanGate leadership employed targeted and personal legal threats as a tactic to suppress the former DMO's concerns. By November 2018, he decided to withdraw his complaint given the lack of progress by OHSA and the high level of harassment he and his family were receiving from active OceanGate legal actions. The claim was closed in December 2018 without OHSA or the Coast Guard having investigated. The resolution of the lawsuit included a non-disclosure clause that was waived by OceanGate following the loss of Titan to allow his participation in the MBI.

4.2 *Director of Engineering (2016–2019)*

OceanGate's Director of Engineering from 2016 to 2019 was an engineer with experience in aerospace and aviation technology, electronics and materials testing. He was hired by OceanGate in 2016 to oversee the integration and testing of the Titan submersible's critical systems. He was initially the only engineer (apart from Mr. Rush), but over time, he built a team of 25. His job title was Director of Engineering (DE), but he testified that the substantive engineering decisions were made by Mr. Rush. The DE's responsibilities included managing the technical aspects of the submersible's development and ensuring that the vessel met performance standards required for deep-sea exploration. He played a key role in the assembly and testing of the Titan I hull, as well as coordinating the manufacturing processes for key components.

The DE testified that he had serious concerns regarding the manufacturing quality of the carbon fiber shell used for the Titan I's hull. He explained that the shell was made from aerospace-grade carbon fiber but was not subject to the same rigorous quality control measures typically used in aerospace applications. He observed inconsistencies in the production process and was concerned about the potential for defects in the shell that could compromise the submersible's structural integrity under deep-sea pressure. A simple example is that the facility where the hull was being fabricated was quite dusty and he asked the manufacturing contractor if they were planning to put a hood over the hull winding machinery to keep the area dust-free. They did not want to do this, and Mr. Rush did not back up the DE's request, so no changes were made.

The DE further testified that the carbon fiber components did not undergo the thorough inspection and testing necessary to ensure they could withstand repeated cycles of extreme pressure. He was concerned that any imperfections introduced during manufacturing would degrade the already razor thin safety margins associated with the immense stress at Titanic depths, leading to catastrophic implosion failure. One significant point of disagreement between the DE and Mr. Rush related to the production of witness panels.[2] The DE thought they were essential for an experimental fabrication with marginal safety factors, but Mr. Rush decided that this effort/expense was not required and so such panels were not manufactured. These and other concerns were documented in writing and were also the subject of repeated arguments with Mr. Rush. As the DE described it:

> Stockton would fight for ... what he wanted, even if it changed from day to day, he wouldn't give an inch much at all. The fights that I had with Stockton were behind closed doors ... most people would eventually back down from Stockton. It was death by a thousand cuts.

As a result, the DE described developing a triage system for his concerns saying:

> If it mattered enough, then I would fight for it. And if it didn't matter enough, I would find a way to mitigate it.

The Titan I hull was damaged during tests in the Bahamas in 2018 and was reluctantly retired in 2019 at the insistence of the DE due to the presence of structural cracking in the carbon fiber pressure vessel hull. In the DE's view, the test program was a success in that the test results demonstrated that the pressure hull vulnerabilities he had warned about were indeed present and the design of Titan I pressure hull did not have the target safety margins for manned dives to the depth of the Titanic. As he described the situation:

> What we're doing has never been done before. We don't have classing [i.e. it's not certified] to tell us what good looks like. Nobody knows what good looks like, but I do know it doesn't look like that.

The DE was dismissed from OceanGate in mid-2019. Mr. Rush told him that the Board was of the view that someone needed to be held accountable for the rupturing of the carbon fiber strands in the hull

[2] Witness panels are extra sections created in parallel with a fabricated item with the same specification specifically designed for testing purposes.

resulting in "hull failure" during testing. Mr. Rush decided that person was the DE.

4.3 Offshore Crew Member (2022)

A contract offshore crew member who worked for OceanGate for only two weeks in July 2022 reached out to the MBI to provide evidence. She has a background in engineering, with qualifications in robotics and experience in marine operations, particularly remote-operated vehicle (ROV) operations in the offshore oil and gas sector.

The crew member was hired as a contractor to work as part of the operations crew for the 2022 season (Titan II) with a promise to be trained as a pilot. Her role initially involved working in mission control as part of the navigation team. She found the approaches used for navigation to be very basic, not up to industry standards, and potentially dangerous. On Dive 79, she reports that for extended periods mission control did not know where the Titan II was relative to the Titanic wreck. She tried to raise concerns and propose better options saying, for example, that in similar applications elsewhere she had seen software-based navigation tracking used instead of manual data handling to eliminate errors and system lag times. This suggestion was not welcomed: she was told she was not "solutions oriented" and was moved from the navigation team into deck operations.

She was onboard during Dive 80 where a loud bang was heard from the Titan II near the end of the dive. This was discussed in the debrief on the morning after the dive but was not seen by Mr. Rush or apparently the rest of the group as a major concern, despite the fact that acoustic monitoring to detect breaking carbon fibers was a key safety system. She described her concern saying: "As an engineer my brain immediately goes to analyzing all possible failure modes". She asked the DE at the time whether this sound could be indicative of delamination in the hull, i.e., the layers of the structure coming apart. He admitted that the strain gauges did show some small movement within the layers of carbon fiber in the hull but that they were going to proceed with the dive program and inspect the hull back in Washington State at the end of dive season.[3]

[3] In fact, Titan remained on the docks in Newfoundland over the winter of 2022/23 because the cost of transporting the vessel back to the OceanGate facilities on the other side of the continent was seen to be unjustified.

Raising concerns was taken as a sign that she was not a good fit for the company. She reports being told “you have a bad attitude … we’re innovative and we’re cowboys and a lot of people can’t handle that” and “many people are concerned about you. You don’t seem to have an explorer mindset”.

She was keen to learn about the technology, but reports that technical questions were not welcome at OceanGate:

> Every question I asked about any of the engineering or the design—the answer was never “here’s the data and here’s the analysis and there’s the testing”. It was always “well this is what Stockton wants” or the foremost expert of whatever said it should be this way … it was sort of name dropping. “NASA does it this way” with no real context. It wasn’t apparent to me what the engineering rigor had been because when you answer the question with “that’s what the company founder wants” rather than design decisions and analysis. That was a red flag to me.

By mutual agreement, she was relieved of her position shortly after this and returned to shore before the end of the diving season. She considered who else she might be able to tell about the unexpected noise on Dive 80 and the general attitude that this would be ignored and concluded that perhaps she should tell the OceanGate Board of Directors. In seeking advice from a friend still with the company as to who the Directors were and how she might be able to contact them, she was effectively talked out of taking any further action on the basis that Mr. Rush was very litigious and she could be opening herself up to being sued.

5 Discussion

5.1 Individual Actions

All three technical professionals who joined OceanGate were drawn by its vision of innovation and its promise of technical excellence. Each was motivated by the opportunity to contribute to a groundbreaking engineering project at the frontier of human exploration and by the company’s early associations with prestigious partners such as the University of Washington’s Applied Physics Laboratory and Boeing. These partnerships suggested that OceanGate operated within the norms of credible, evidence-based engineering practice—something that carries significant weight for technical professionals who value competence, rigor, and adherence to sound engineering standards (Middlehurst & Kennie, 1997). Yet once inside the organization, they discovered that these affiliations were short-lived and largely symbolic. As the company distanced itself from these institutions, its credibility among technically minded staff was eroded.

All three professionals quickly developed concerns about aspects of OceanGate's activities within their given areas of responsibility. They adopted similar strategies to communicate these concerns to Mr. Rush with a view to preventing future serious problems. In this spirit, the DMO suggested that testing and certification were necessary to ensure hull integrity and crew safety based on his experience with other submersibles. The DE wanted improvements to the hull manufacturing process to reduce the chances of later hull integrity issues. Similarly, the offshore crew member proposed improvements to navigation processes to improve crew safety. All individuals identified specific risks, proposed practical mitigation steps, and communicated their reasoning based on evidence and prior experience. Yet their recommendations were dismissed as unnecessary delays that conflicted with the CEO's aggressive development timeline.

The differences in attitudes toward failures between these technical experts and OceanGate management are also evident. The DE noted that the damage to the Titan I pressure hull during testing in 2018 demonstrated the success of the testing program of this innovative technology, i.e., it provided useful data so that the design could be improved in the next version of the hull. Taking a similar orientation toward data potentially signaling a problem, the offshore crew member saw the sounds heard from Dive 80 as a possible indication of delamination that required immediate investigation. Both individuals saw failures as providing useful evidence of system performance that should be investigated and used for learning. In contrast, OceanGate management chose to blame the DE for the 2018 failure and, some years later, ostracize the offshore crew member when she sought to highlight the Dive 80 issues. OceanGate management continued to see failures as obstacles to progress to be ignored, minimized, or blamed on others rather than vital warnings of an unsafe hull design. They apparently had no sense of the "gift of failure" (Carroll & Fahlbruch, 2011).

5.2 *Moving from Insider to Outsider*

We see these technical professionals enacting many of the behaviors that the literature suggests would be effective in convincing senior management that their advice should be acted upon such as drawing on failures for learning, linking recommendations to potential consequences but also to business objectives, providing solutions and documenting their reasoning. Yet in each case, their advocacy for change was rejected. For those involved, this created a profound ethical dilemma—how to honor their professional responsibility to protect public safety when their voices were ignored within the organization.

At one level, the reason they were unsuccessful lies in their assumption that the CEO's actions were based on ignorance and so their input would be welcomed as new technical information and seen as an opportunity for organizational learning (Weick & Sutcliffe, 2015; Weick et al., 1999). In fact, the actions of the CEO were not grounded in considerations of technical expertise but rather drawn from in a different set of priorities including valuing innovation for its own sake and needing to generate revenue from paying passengers. It may be tempting to conclude that individuals cannot work for organizations with differing values and that uncovering such differences must lead to experts departing from an employing organization. But this is an oversimplification.

We know that organizations with high employee embeddedness have lower turnover. That is, the more strands connecting an employee socially, psychologically, and financially to a given organization the more bound they are to their job and their organization (Maurer & Qureshi, 2019). In the Titan case, as a result of the links to other prestigious engineering bodies as described above, our three experts joined OceanGate expecting to become part of a group that valued technical excellence, but they were disappointed. This form of embeddedness did not materialize. As a result of speaking up for safety, all also became socially isolated at work and ultimately were pushed from being an insider to an organizational outsider. The DMO produced a comprehensive report detailing his concerns and effectively refused to accept handover of the vessel into operations which initiated face-to-face conflict with Mr. Rush, resulting in his departure from the organization. The DE tried to use soft power to influence Mr. Rush's decision making, but when things went wrong, he found himself held responsible and excluded from the organization by being dismissed. The offshore crew member tried to raise concerns about the noise heard in Dive 80 but was ostracized as a result and pushed out of the organization. Social isolation changed the possible options available for them to advocate change. As advocacy on safety generated increasingly more hostile organizational responses, they were left with no choice except to conform and fall into line, or to escalate their concerns to the degree that they risked further damage to already strained social relationships.

The size of OceanGate amplified the significance of these social relations. Mr. Rush was not a largely faceless CEO in another city but someone that the technical professionals interacted with every day. They knew each other. The disagreements were a constant interpersonal struggle. Whistleblowers generally leave organizations reluctantly, disappointed that their colleagues, as well as the organization, have let them down. Indeed, it can be those very social relationships that make the move to becoming an outsider so difficult. We see this in the story of John Kiriakou, former CIA officer who blew the whistle about use of torture as official US policy in the "War on Terror". He refused to be trained in enhanced interrogation techniques and explains (Kiriakou, 2025, p. 35):

> The reason that's difficult for me, even after all these years, is that many of the other men who did go through the training and who did carry out the torture were friends of mine. Our children played together. We went on holiday together. We had barbeques on the weekends. I liked these men. I had no idea they had it in them to carry out something as heinous as a torture program.

When it comes to attitudes to safety, this suggests that the crunch comes when the conflict is not a disagreement with nameless, faceless managers but when you think the person sitting next to you is doing the wrong thing. Values are important, but it is this personal component to value conflict that makes it untenable for employees to remain.

Our analysis of the impact of social embeddedness goes well beyond the dynamics of power imbalances that are the usual focus of academic attention in this context. The dual consideration of the level of social embeddedness and the orientation toward value-based decision making supports four possible behavioral archetypes about individual engineering behaviors for speaking up as shown in Fig. 6.3.

The dilemma is that social embeddedness seems likely to amplify values orientation and associated behaviors. Values-driven engineers with high social embeddedness are likely to put the most effort into speaking up for

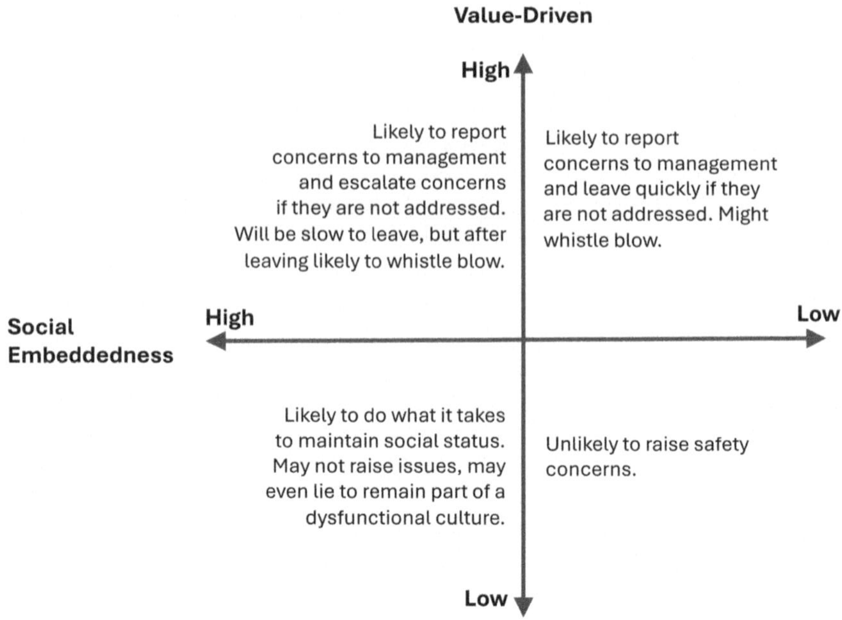

Fig. 6.3 The impact of social embeddedness and values orientation on engineering speaking up behaviors

safety. Even if their initial concerns are dismissed, they are likely to continue trying to speak up and will be slow to walk away. If they feel they are forced to leave an organization and become an outsider, their ongoing concerns are most likely to lead this group to become whistleblowers. In contrast, high social embeddedness combined with a lower concern for professional ethics may lead to lying behaviors, simply to get along. In the extreme, members of this archetype are likely to leave an organization and move to other employment, without blowing the whistle.

Low social embeddedness and highly values-driven engineers are likely to speak up with few social constraints although they have little to keep them attached to an employer and may quickly move on if their concerns are not addressed. Members of the archetype with low social embeddedness and low concern for values are unlikely to speak up for safety.

5.3 *Is Whistleblowing the Answer?*

In the Titan case, the engineers all moved from insider to outsider status, but they differed in some aspects of their responses to the constant conflict. The DE chose to remain in the organization longest and try to influence the CEO on the matters he considered to be most important. This choice to compromise, which he described as "death by a thousand cuts", allowed him to remain working at OceanGate despite the misalignment in professional values. In the end, his efforts to influence the organization cannot be seen as successful given that he was dismissed after being blamed for engineering problems he himself had anticipated.

The shift from offering technical advice to formally reporting one's employer to external authorities is enormous, requiring courage and clarity about both the risks and responsibilities involved. The DMO, however, chose to take that step. By failing to evaluate the engineering warnings embedded in his report, regulators missed a crucial opportunity to intervene before tragedy struck. His actions, at significant personal cost, were also not successful in preventing the disaster. As she departed OceanGate, the offshore crew member also considered some form of whistleblowing such as contacting the Board and reporting her concerns. In the end, she chose not to, aware of the actions taken against the DMO a few years earlier and concerned at the personal cost she may incur because of any whistleblowing activities she might choose to initiate.

The Titan case shows that whistleblower systems have a long way to go to be effective in either preventing adverse events or protecting those who seek change (Munro et al., 2025).

6 Conclusion

Engineers and technical professionals do not act in isolation; they operate within the social and organizational systems that create and sustain technology. Yet within those systems, there remains significant individual agency—the capacity to act professionally, ethically, and constructively in the interest of safety. Sometimes there comes a point when ethical professionals become aware that their values are not shared with decision makers in their organization. This can result in significant conflict in the form of robust dialog. Conflict grounded in strong argument (rather than personal attack) is not necessarily a bad thing for safety. In the best case, it results in changing hearts and minds. Even if this is not the outcome, preserving one's professional integrity and continuing to advocate for safety is critical, especially if we consider that the alternative is to back down and become part of the problem, as we can see in cases such as the collective lying that can grow in organizations when the truth is completely unpalatable to management (Hayes et al., 2024).

Decision makers should be aware that employees who are values-driven and strongly embedded in an organization will put more effort into trying to get action on safety. Their organizational insider status is of value to them as well as their sense of professional ethics. They will seek to align these and have their safety concerns acted upon, potentially for an extended period before they make the choice to back down or change behaviors and move to becoming an organizational outsider. Organizations should take professionals' concerns seriously knowing that they come from a sense of wanting to improve organizational performance, rather than intending harm.

Equally, ethical professionals who find their concerns ignored should be wary of using their sense of organizational embeddedness as an excuse for compromising in a major way on professional ethics. Where a professional feels the only ethical paths available involving leaving an organization, they will need to grapple with whether to become whistleblowers or go quietly. Both options carry significant personal and professional consequences. Yet both can serve an ethical purpose. Leaving on principled grounds signals dissent from unsafe practices; whistleblowing, when done responsibly, preserves vital safety knowledge that might otherwise be lost.

While our focus in this chapter has been the strategies and actions of technical professionals, we must emphasize the critical role of organizational and governance protocols in supporting ethical action on the part of individuals. The Titan failure should not be viewed as a failure on the part of the technical professionals whose views were ignored. It was in very large part due to the values and direct actions of the CEO. When technical professionals are placed in a situation to contradict the values/beliefs of superiors, they proceed with great personal risk. There exist layers of governance that can serve as a "check" on the decision making of the CEO including internal parties like OceanGate's Board of Directors as well as external entities such as industry regulators. In the case of Titan, the Board played no active role in ensuring the integrity of the submersible and the safety of its crew. OSHA, the regulator, also failed to responsibly investigate the safety concerns that were brought to their attention. Boards and regulators have a critical role to play in supporting values-driven engineers in the spirit of enhancing safety.

The analysis of the Titan case also suggests future research directions. Beyond individual accident or whistleblowing case studies, the reciprocal nature of safety communications has received little scholarly attention with research focusing almost exclusively on workers as senders of a message. We echo other appeals that research on the expert/manager dyad is needed (Paul et al., 2025). Regarding individual professionals, accident prevention efforts would benefit from a greater understanding of how the social nature of their work interacts with ethical considerations and can both encourage and constrain speaking up for safety.

References

Aven, T. (2009). Perspectives on risk in a decision-making context—Review and discussion. *Safety Science, 47*(6), 798–806. https://doi.org/10.1016/j.ssci.2008.10.008

Carroll, J. S., & Fahlbruch, B. (2011). 'The gift of failure: New approaches to analyzing and learning from events and near-misses.' Honoring the contributions of Bernhard Wilpert. *Safety Science, 49*, 1–4. https://doi.org/10.1016/j.ssci.2010.03.005

Columbia Accident Investigation Board (CAIB). (2003). *Report of the Columbia Accident Investigation Board: Volume 1.* U.S. Government Printing Office.

Committee on Transportation and Infrastructure. (2020). *Final Committee Report: The design, development and certification of the Boeing 737 MAX.* U.S. House of Representatives.

Gonzalez, J. A., & Pérez-Floriano, L. R. (2015). If you can't take the heat: Cultural beliefs about questionable conduct, stigma, punishment, and withdrawal among Mexican police officers. *Organization Studies, 36*(5), 665–687. https://doi.org/10.1177/0170840615571961.

Hayes, J. (2015). Taking responsibility for public safety: How engineers seek to minimise disaster incubation in design of hazardous facilities. *Safety Science, 77*, 48–56.

Hayes, J., & Maslen, S. (2022). Putting senior management in focus: A critical reflection on the direction of safety science research. In K. P. Gould & C. Macrae (Eds.), *Inside hazardous technological systems: Methodological foundations, challenges and future directions.* CRC Press.

Hayes, J., & Maslen, S. (2023). Preventing major disasters: Success and failure as two sides of the same coin. In A. Mica, M. Pawlak, A. Horolets, & P. Kubicki (Eds.), *Routledge international handbook of failure.* Routledge. https://doi.org/10.4324/9780429355950-24

Hayes, J., Maslen, S., Scott-Young, C., & Wong, J. (2017). The rise of defensive engineering: How personal liability considerations impact decision-making. *Journal of Risk Research, 21*(10), 1293–1307. https://doi.org/10.1080/13669877.2017.1391319

Hayes, J., Maslen, S., Holdsworth, S., & Sandri, O. (2021). Defining the capable engineer: Non-technical skills that support safe decisions in uncertain, dynamic situations. *Safety Science, 141*, 105324. https://doi.org/10.1016/j.ssci.2021.105324

Hayes, J., Maslen, S., & Schulman, P. (2024). A case of collective lying: How deceit becomes entrenched in organizational safety behavior. *Safety Science, 176*, 106554. https://doi.org/10.1016/j.ssci.2024.106554

Hayes, J., Maslen, S., Holdsworth, S., & Sandri, O. (2025). The part that we play: Engineers' perceptions of their responsibility for public safety. *Safety Science, 187*, 106862. https://doi.org/10.1016/j.ssci.2025.106862.

Hopkins, A. (2019). *Organising for safety: How structure creates culture*. CCH.

Hopkins, A. (2025). *Boeing, the 737 MAX crisis and aviation safety*. CRC Press. https://doi.org/10.1201/9781003629269

Johnson, D. G. (2020). *Engineering ethics: Contemporary and enduring debates*. Yale University Press.

Jones-Lee, M., & Aven, T. (2011). ALARP—What does it really mean? *Reliability Engineering and System Safety, 96*, 877–882. https://doi.org/10.1016/j.ress.2011.02.006

Kiriakou, J. (2025). 'Why we all have to fight': An insider speaks out on the crimes of the CIA. In I. Munro, K. Kenny, & M. Fotaki (Eds.), *Perspectives on whistleblowing: Cases and theories*. Bristol University Press.

Klein, G. (1999). *Sources of power: How people make decisions*. MIT Press.

Maslen, S., Hayes, J., Holdsworth, S., & Sandri, O. (2021). When ethics is a technical matter: Engineers' strategic appeal to ethical considerations in advocating for system integrity. *Science and Engineering Ethics, 27*(4), 1–20. https://doi.org/10.1007/s11948-021-00324-7.

Maurer, C. C., & Qureshi, I. (2019). Not just good for her: A temporal analysis of the dynamic relationship between representation of women and collective employee turnover. *Organization Studies, 42*(1), 85–107. https://doi.org/10.1177/0170840619875480

Middlehurst, R., & Kennie, T. (1997). Leading professionals: Towards new concepts of professionalism. In J. Broadbent, M. Dietrich, & J. Roberts (Eds.), *The end of the professions? The restructuring of professional work*. Routledge.

Motarjemi, Y. (2023). Whistleblowing: An essential element of public health and food safety management. In *Food safety management: A practical guide for the food industry* (2nd ed., pp. 1089–1107). https://doi.org/10.1016/B978-0-12-820013-1.00025-5

Munro, I., Kenny, K., & Fotaki, M. (Eds.). (2025). *Perspectives on whistleblowing: Cases and theories*. Bristol University Press.

Paul, J., Pilbeam, C., & Smallwood, A. (2025). Safety voice concept clean-up: Examining the voice that challenges us to be safer. *Safety Science, 191*, 106931. https://doi.org/10.1016/j.ssci.2025.106931

Pidgeon, N., & O'Leary, M. (2000). Man-made disasters: Why technology and organizations (sometimes) fail. *Safety Science, 34*, 15–30.

Reader, T. W., & O'Connor, P. (2014). The deepwater horizon explosion: Nontechnical skills, safety culture, and system complexity. *Journal of Risk Research, 17*(3), 405–424. https://doi.org/10.1080/13669877.2013.815652

Reason, J. (1997). *Managing the risks of organizational accidents*. Ashgate.

Rogers, W. (1986). *Report to the president by the presidential commission on the space shuttle challenger accident*. NASA.

Schmidt, C. (2020). Why risk management frameworks fail to prevent wrongdoing. *Learning Organization, 27*(2), 133–145. https://doi.org/10.1108/TLO-10-2019-0150

Starbuck, W. H., & Farjoun, M. (Eds.). (2005). *Organization at the limit: Lessons from the Columbia disaster*. Blackwell Publishing.

Stene, L. K., & Eikemo, I. (2020). Is whistleblowing a safety mechanism or a threat? A study of petroleum-related companies in Norway. In *Proceedings of the 29th European Safety and Reliability Conference (ESREL 2019)*. Research Publishing.

van de Poel, I., Fahlquist, J. N., Doorn, N., Zwart, S., & Royakkers, L. (2012). The problem of many hands: Climate change as an example. *Science and Engineering Ethics, 18*, 49–67. https://doi.org/10.1007/s11948-011-9276-0

Vaughan, D. (1996). *The challenger launch decision: Risky technology, culture and deviance at NASA*. University of Chicago Press.

Weick, K. E., & Sutcliffe, K. M. (2001). *Managing the unexpected: Assuring high performance in an age of complexity*. Jossey-Bass.

Weick, K. E., & Sutcliffe, K. M. (2015). *Managing the unexpected: Sustained performance in a complex world*. Wiley.

Weick, K. E., Sutcliffe, K. M., & Obstfeld, D. (1999). Organizing for high reliability: Processes of collective mindfulness. In R. I. Sutton & B. M. Staw (Eds.), *Research in organizational behavior* (Vol. 21, pp. 81–123). JAI Press.

Whalen, T. F., & Neubauer, J. D. (2025). *Report of the Marine Board of Investigation into the Implosion of the Submersible Titan (CG1788361) in the North Atlantic Ocean near the Wreck Site of the RMS Titanic resulting in the loss of five lives on June 18, 2023*. U.S. Coast Guard.

Whitmore, J. L., & Keeler, J. B. (2025). Dare to share: A path analysis and social exchange investigation of workplace social courage and safety knowledge sharing. *Safety Science, 191*, 106973. https://doi.org/10.1016/j.ssci.2025.106973

CHAPTER 7

What Does It Take to "Settle" a Scandal? The Ethics of Hearing and Acting on Safety Concerns

Dawn Goodwin and Daniel Taylor

1 Introduction

The Report of the Gosport Independent Panel (Jones, 2018) documents an inquiry into the overuse of diamorphine on an elderly care unit, between 1987 and 2001, and the subsequent struggle to bring this issue to light. Diamorphine (a strong opiate painkiller), administered via a syringe-driver, was prescribed routinely and inappropriately, and is estimated to have shortened the lives of 450 patients. Nurses first raised concerns in 1991 but were effectively silenced by dismissive senior healthcare staff. However, concerns continued to emerge in the form of patient complaints to the hospital, the police, and healthcare regulators. A series of investigations and inquests ensued, but the death rate on the unit only returned to

D. Goodwin (✉)
Lancaster University, Lancaster, UK
e-mail: d.s.goodwin@lancaster.ac.uk

D. Taylor
University of Sheffield, Sheffield, UK
e-mail: daniel.taylor@sheffield.ac.uk

J. Hayes, S. Maslen (eds.), *Ethics in the doing of technoscience*,
https://doi.org/10.1007/978-3-032-20002-0_7

normal when the doctor primarily responsible for clinical care resigned in 2000. The doctor continued to work as a GP and in 2010 was found guilty of serious professional misconduct. She retired and relinquished her General Medical Council (GMC) registration. But still, the scandal persisted and in 2014 an independent inquiry was commissioned. Even after the inquiry report was published in 2018, discontent continued and the police commenced fresh investigations—at the time of writing in May 2025 they were still ongoing (Kent Police, 2025).

We analyze this case to explore the ethics of hearing and acting on safety concerns. We focus on the circumstances in which concerns about safety are either heard or silenced, and where they are acted upon, whether that action satisfied the concern-raiser. To help us understand these dynamics, we draw on scholarship around epistemic injustice which guides us to scrutinize how prejudice, stereotyping, and discrimination infiltrate judgments about the credibility of speakers as well as working at a structural level to constrain or discredit certain voices.

The chapter proceeds as follows: we first describe how we have used the inquiry report as a source of data for our analysis and introduce the concepts we employ. We then focus on the nurses and families who were subject to enduring epistemic injustice. In doing so, we examine the contexts and processes of knowledge production, specifically the handling of concerns and complaints by hospital management and subsequent police investigations, to explore how epistemic oppression is perpetuated through agents and structures. Our analysis addresses how ethical matters are expressed and muted, and how epistemic injustice impacts decision making around raising, hearing or dismissing ethical concerns about healthcare practice. We conclude by exploring the implications of this case for contemporary ethical practice in healthcare.

2 Methods and Conceptual Resources

Inquiry reports are widely considered to produce the definitive version of events; however, Boudes and Laroche (2009) caution against positioning the reports as unbiased windows on what happened. They highlight how a seemingly objective chronology entails choices that suggest a storyline and the inclusion of some occurrences and not others may cast the events in a different light. Further, the insertion of comments and the way data is presented informs how evidence should be interpreted (Boudes & Laroche, 2009; Goodwin, 2018). Therefore, we acknowledge that The Report of the Gosport Independent Panel gives a specific view of events and filters information to produce a coherent narrative that speaks to a specific purpose.

Inquiries are often commissioned to respond to public suspicion that information is being withheld, and Gosport was no exception. The Independent Panel was tasked with undertaking maximum disclosure of relevant information relating to unexpected deaths at Gosport War Memorial Hospital and to produce a report which provided an account of events. In doing so, the report guides the reader through many complex issues, immense amounts of information and situates events in historical and cultural context. We are, therefore, mindful of the panel's editorial role, in that inclusions, exclusions, and contextualizations are at the command of the inquiry team. We make no claim that this is the only true story that can be told about events at Gosport, and do not see it as a complete record. However, The Report of the Gosport Independent Panel is thoroughly researched with every document catalogued, referenced, and available for public scrutiny on the online archive [https://www.gosportpanel.independent.gov.uk/document-library/online-archive/]. We therefore position the report as a verifiable and comprehensive account, but nevertheless, one that responds to certain public and political demands. Furthermore, our analysis of the report has been parsimonious; we have focused only on the chapters that illuminate issues of epistemic injustice when raising patient safety concerns, namely, Chap. 1 (the nurses' concerns), Chap. 3 (experiences of patients and families), and Chap. 5 (the police investigations and Crown Prosecution Service), providing a summary of the information covered in these chapters and a narrative that explores issues of epistemic injustice.

Theoretically, we draw on epistemic injustice as we believe it captures something important about what goes wrong when complaints are made and dismissed. Fricker (2007) defines two types of epistemic injustice: testimonial injustice which "occurs when prejudice causes a hearer to give a deflated level of credibility to a speaker's word" (Fricker, 2007, p. 1), and hermeneutical injustice, which refers to a situation where the concepts for properly understanding one's experience do not exist, putting them at a disadvantage. There has, however, been considerable debate about Fricker's account of epistemic injustice, particularly that what Fricker describes as "epistemic bad luck" might be more properly understood as structural oppression of certain patterns of experience (e.g., see Berenstain, 2020). We therefore draw on feminist decolonial scholarship to elucidate the more structural elements of epistemic injustice.

3 Epistemic Exclusions: Silencing Nurses' Concerns

Box Example 1

Early in 1991, a staff nurse on an elderly care ward, Anita Tubbritt, contacted the Royal College of Nursing (RCN) to express concerns about the overuse of diamorphine and its administration via syringe-driver. After meeting with five or six members of staff, the RCN representative, Keith Murray, suggested the nurses' concerns be escalated to Mrs. Isobel Evans (the Patient Care Manager at the hospital).

A meeting between Mrs. Evans, staff nurse Giffin (one of the nurses involved in the initial meeting) and Mr. Murray took place in April. It was decided to display a notice stating the RCN were aware of the concerns and to arrange a meeting, for all staff, where concerns would be heard without fear of reprisals or disciplinary action, and that a written policy on syringe-drivers and controlled drugs would be agreed.

The staff meeting was attended by Mrs. Evans, eight nurses and two ward sisters. The following concerns were discussed:

1. Not all patients given diamorphine had pain.
2. Other forms of analgesia were not considered, and the "sliding scale" for analgesia was never used.
3. The drug regime was used indiscriminately, patients' individual needs were not considered, nor was oral and rectal treatment ever considered.
4. Patients' deaths were sometimes hastened.
5. The use of syringe-drivers prohibited trained staff from adjusting the dose to suit patients' needs.
6. Levels of sedation resulted in too high a degree of unresponsiveness.
7. Sedative drugs such as Thioridazine would sometimes be more appropriate.
8. Diamorphine was prescribed prior to procedures such as catheterization where diazepam would be as effective.

9. Staff's views were not considered before starting patients on diamorphine.
10. Similar units did not use diamorphine as extensively.

Mrs. Evans responded that patients suffered distress from symptoms besides pain but also had the right to a peaceful and dignified death, and that most patients had complex problems. Nevertheless, she would invite the nurse manager for elderly services at another hospital to discuss with staff the issues they raised.

To improve her knowledge, staff nurse Tubbritt attended a course on elderly care at which she voiced concerns about the use of syringe-drivers, the lack of training for those administering the drugs and initiated a class discussion on the "use and abuse of the syringe-driver". The course tutor subsequently visited the ward in October 1991 to meet the other nurses. The nurses reiterated their concerns, adding that despite raising their concerns with their manager in July, no action had been taken. The tutor circulated a report of the meeting to those who attended and a number of others including Mrs. Evans and Mr. Bill Hooper (the General Manager for the hospital).

Nurse Tubbritt again contacted Mrs. Evans, but this time Mrs. Evans' response was more defensive, referring to "disruptive criticism which achieves nothing positive and leaves staff feeling frustrated". Shortly after the tutor's visit, Mrs. Evans wrote to all trained staff, copying her letter to the Night Sister, Dr. Logan (Consultant Geriatrician), Dr. Barton (Clinical Assistant), and Mr. Hooper. In this, she referred to staff concerns as "allegations" and requested the names of patients they believed had been prescribed drugs inappropriately. Mr. Murray (the RCN representative) then escalated the issue to the District General Manager of the Health Authority requesting advice, given that the nurses' concerns had been dismissed as "a small group of night staff who are 'making waves'".

With no response to her letter, Mrs. Evans arranged a meeting in December for all staff about the use of diamorphine. It was attended by only four nurses this time, alongside a ward sister, Mrs. Evans and Drs. Logan and Barton. In this meeting, the nurses were cautioned to keep their comments objective, instructed to keep further concerns internal, and were told they had made the doctors feel "very threatened and defensive". Mrs. Evans emphasized the stress the issue had put on the medical staff and warned that it had "the potential of being detrimental to patient care and relatives' peace of mind and could undermine the good work being done in the unit if allowed to get out of hand". The nurses were given the impression that because they were night staff, they did not see the whole picture and their concerns reflected a lack of communication with day staff. The meeting concluded with Mrs. Evans asking if a policy on diamorphine and syringe-drivers was needed, to which no one responded. No further action was taken by the hospital.

Testimonial injustice occurs "when someone is wronged in their capacity as a giver of knowledge" (Fricker, 2007, p. 3). A central feature of this is how prejudice causes the hearer to give the speaker less credibility than they would otherwise have given (Fricker, 2007). Fricker contends that assessments of a speaker's credibility are ordinarily informed by stereotypes. She explains that "much of everyday testimony requires the hearer to engage in social categorization of speakers" as "barring a wealth of personal knowledge of the speaker as an individual, such a judgment of credibility must reflect some kind of social generalization about the epistemic trustworthiness—the competence and sincerity—of people of the speaker's social type, so that it is inevitable … that the hearer should spontaneously avail himself of the relevant generalizations in the shorthand form of … stereotypes" (Fricker, 2007, p. 32). Prejudice works through stereotypes by associating certain (typically less powerful) groups with attributes inversely related to competence or sincerity such as over-emotionality, illogicality, inferior intelligence, lack of moral fiber, and so forth (Fricker, 2007).

In handling the nurses' concerns, there are several ways in which testimonial injustice is incurred. First, while initially appearing to hear the nurses' concerns, Mrs. Evans engaged in a form of ethical gatekeeping, shaping which concerns were considered legitimate and which were dismissed. She cast doubt on the nurses' assessment, noting the complexity of patients' needs and implicitly questioning whether the nurses' assessment was sufficiently nuanced. As the nurses persisted, the reframing of concerns became more overt; they were branded as "allegations" and "disruptive criticism" that made doctors defensive and threatened the functioning of the unit. As Fricker puts it, a "credibility deficit" is at work when the cause of the nurses' (simplistic) judgments arises from their position as night nurses unable to see the full picture. In all these ways, the nurses' competence is undermined. Fricker points out, however, that "epistemic trustworthiness has two distinct components: competence and sincerity" (Fricker, 2007, p. 45), and doubt was also cast on the nurses' motives when they were described as a small group of night staff intent on "making waves". Further, framing the issue as stressful for doctors rather than dangerous for patients inverted the ethical landscape, prioritizing professional comfort over patient welfare.

All this points toward the conclusion that the nurses' views on the use of diamorphine via syringe-drivers was dismissed by virtue of *who* they were; their social identity as a group of night nurses—and so marginal even within nursing ranks—placed them at a disadvantage when, in the hierarchical context of 1990s healthcare, they challenged the practice of a doctor supported by the consultant geriatrician and hospital management. The nurses made considerable efforts to resist having their assessment undermined; nurse Tubbritt undertook a course on care of the elderly, and they attempted to work collectively, drawing strength from their shared perception of the inappropriateness of prescribing habits and thus avoiding the suggestion that an individual's judgment was faulty. Further, they engaged informed external viewpoints such as the RCN representative and course tutor. However, in a context where competence is aligned with rank, these efforts were insufficient.

But, as Anderson (2012) suggests, the roots to testimonial injustice go beyond the hierarchies of professional identities, they are embedded in organizational structures. First, the informal routes through which the nurses raised their concerns were shut down with stipulations that concerns must be in writing and contain specific patient details. This shift transformed an ethical issue of patient safety into a bureaucratic hurdle, effectively discouraging further action. Thus, a general concern about routine practices on the ward became a formal complaint about the prescribing habits of a particular doctor. The demand for specific patient details also allows for dispute in medical terms, in which doctors are better equipped than nurses to argue their case. The nurses were instructed to keep concerns internal, thereby isolating them from external expertise. Anderson argues that "testimonial exclusion becomes structural when institutions are set up to exclude people without anyone having to decide to do so" (Anderson, 2012, p. 166), and here the structure of the December meeting was made sufficiently intimidating as to deter all but four nurses, reduce the collective voice, silence any further concerns, and close down the possibility of a formal policy on diamorphine via syringe-drivers.

4 Asymmetries in Epistemic Resources: "Palliative Care" and Communication with Families

Box Example 2

Gladys Richards was a 91-year-old nursing home resident with dementia. She was frail, incontinent, had hearing difficulties, and required help with washing and dressing. Mrs. Richards' daughters and granddaughter visited daily and were heavily involved in her care. After a six-month history of falls at the nursing home, on July 29, 1998, Mrs. Richards fell and fractured her right neck of femur which required surgical intervention. Afterward, she had little discomfort and could comfortably sit in a chair. On August 11 she was transferred to Gosport War Memorial Hospital for rehabilitation. At this point, Mrs. Richards was unable to bear weight fully but could walk with assistance. Dr. Barton saw Mrs. Richards on admission, noting she was frail and wrote in the records: "not obviously in pain please make comfortable ... I am happy for nursing staff to confirm death". Dr. Barton prescribed morphine, diamorphine, hyoscine (an anti-sickness drug), and midazolam (a sedative) to be administered by syringe-driver over 24 hours. On August 13, Mrs. Richards fell and dislocated her hip, requiring intervention at Haslar Hospital before returning to Gosport War Memorial Hospital on August 17. Haslar Hospital advised that Mrs. Richards could mobilize and would need no further follow-up. This time Mrs. Richards arrived at Gosport War Memorial Hospital in pain and distress. A syringe-driver was commenced, delivering a continuous infusion of potent drugs which were administered until her death on August 21.

Mrs. Richards' daughters were unhappy about the care of their mother. The day before Mrs. Richards' death they complained, verbally, to the Director of Quality at the Portsmouth Healthcare NHS Trust about poor standards of care. They complained further after her death, prompting an internal investigation. While this investigation identified the syringe-driver as an element of care the family were unhappy with—its use had left Mrs. Richards unconscious and family members unable to say "goodbye"—its contents and their potent sedative effects were not acknowledged. It would appear Mrs. Richards' care was discussed with her daughters, but the potency of the syringe-drivers' contents was not fully comprehended. Further, Mr. Hooper (the General Manager at the hospital), who commissioned the internal report into Mrs. Richards death, failed to identify any link between Mrs. Richards' care and the nurses' concerns over opioid use seven years previously.

Box Example 3

Arthur Cunningham was a 79-year-old nursing home resident who suffered from multiple health conditions, including Parkinson's disease, dementia, diabetes, reactive depression, and myelodysplasia (a bone marrow disorder). Mr. Cunningham was admitted to Gosport War Memorial Hospital on September 21, 1998, for treatment of a large sacral pressure sore. His nursing home was asked to hold his place for up to three weeks as he was expected to return.

Dr. Lord (Consultant Geriatrician) noted "prognosis poor" and prescribed oral morphine to be administered prior to dressing his sacral sore. Dr. Barton later assessed Mr. Cunningham and recorded in the notes "make comfortable. Give adequate analgesia. I am happy for nursing staff to confirm death" and prescribed a syringe-driver containing diamorphine, midazolam, and hyoscine. From this point, there were two instances of communication between hospital staff and Mr. Cunningham's son: on September 22, the nursing notes recorded "[son] has telephoned. Explained that syringe-driver containing diamorphine, midazolam was commenced yesterday evening for pain relief and to allay his anxiety", and the following day, the nursing notes record that Mr. Cunningham "has become chesty overnight, to have hyoscine added to driver ... son informed of deterioration ... asked if this was due to the commencement of the syringe-driver ... informed that [Mr. Cunningham] was on small dosage which he needed".

It seems these two conversations obfuscated Mr. Cunningham's clinical condition as the clinical notes subsequently recorded that Mr. Cunningham's son was "very angry that syringe-driver had been commenced. It was explained yet again that the contents of the syringe-driver were to control his pain". It was also explained that it would need the consultant's permission to discontinue the driver and prescribe an alternative method of pain relief. The entry ends "now fully aware that [Mr. Cunningham] is dying and needs to be made comfortable". Further entries in the clinical documentation show that doses of sedative medication were tripled in single instances, without adequate documentation that this was explained properly to Mr. Cunningham's son. Mr. Cunningham died on the evening of September 26.

Hermeneutical injustice exists "when a gap in collective interpretive resources puts someone at an unfair disadvantage when it comes to making sense of their social experiences" (Fricker, 2007, p. 1). On Fricker's account, hermeneutical injustice is not perpetrated by agents, it is a structural issue—an absence of the necessary concepts for sense-making. Dotson (2012) points out two problems with this formulation: (1) often the hermeneutical resources do exist and are well-used by marginalized communities to account for their experiences; but (2) the problem arises when such concepts are not recognized by powerful others who maintain a position of "willful hermeneutical ignorance". Dotson proposed the term "contributory injustice" to describe circumstances where "an epistemic agent's willful hermeneutical ignorance in maintaining and utilizing structurally prejudiced hermeneutical resources thwarts a knower's ability to contribute to shared epistemic resources within a given epistemic community by compromising her epistemic agency" (Dotson, 2012, p. 32). This identifies how hermeneutical resources may exist, but there are asymmetries in accessing and deploying them. While, in the cases we outline, the asymmetries do not work in the same way as Dotson describes, her discussion of contributory injustice is helpful in focusing attention on unequal access to hermeneutical resources.

We have given two examples—many more are detailed in the report—where patients were admitted for active treatment or rehabilitation, yet they were effectively given palliative care with the indiscriminate practice of prescribing a combination of drugs frequently reserved for treating intractable symptoms as patients near the end of their lives. The outcome of this course of treatment was well understood by Dr. Barton and the nursing staff, being clearly indicated by the routine statement "Happy for nursing staff to confirm death". However, these decisions rarely involved adequate input from patients or their families, and there were numerous occasions where these high-consequence clinical decisions were not properly communicated.

The cases above illustrate how patients and their families were excluded from treatment decisions. Patients' and families' status as knowers and their ability to make informed judgments were dependent upon the information given to them by nurses and doctors. However, insufficient and ambiguous information from clinical staff prevented many from comprehending the clinical trajectory. Our argument is that the doctors and nurses were agents of contributory injustice, which Dotson defines as when "an epistemic agent … obscures or distorts genuine epistemic differences through the use of a different, incompatible set of hermeneutical resources" (Dotson, 2012, p. 38). When questioning the syringe-driver and its contents, staff responded in terms of soothing pain and agitation. Only rarely did explanations involve explicit reference to end-of-life. In Mr. Cunningham's case, it took his son to forcefully challenge the need for the syringe-driver for the nurses to acknowledge that he was dying. It should be noted that neither of the patients discussed above were admitted for palliative care, but this was clearly the guiding concept and framework within which the doctors and nurses were operating. This communication failure—obscuring the nature of medications being administered—rendered shared decision making with patients and family members impossible, and constituted negligence in the professionals' ethical duty to ensure informed consent and respect patient autonomy and dignity.

As Carel and Kidd (2018) explain, the agential/structural distinction is complicated in healthcare, involving individual agency and structural features of healthcare systems such as time pressure, routinization of care, and formalization of diagnostic practices. At Gosport, structural biases that legitimized withholding conceptual resources from patients and families were compounded by institutionalized beliefs about goals of care which were at odds with best practice at the time (e.g., decisions to commence pre-emptive medication via syringe-driver). Principally, the medical profession's monopoly over knowledge of health and treatment meant that families had little access to alternative explanations for their loved ones' care. There was a strong cultural respect for authority among the families which made it even more difficult for the families to challenge the explanations they were given (Jones, 2018). This imbrication of agential-structural barriers thwarts families' efforts to question the care provided and is indicative of a broader issue within healthcare institutions, where power imbalances between medical professionals and families inhibit ethical decision making. Carel and Kidd (2018, p. 336) explain that healthcare relies on:

> complex structures of epistemic norms and expectations, both implicit and explicit, that create knowledge asymmetries—for instance, privileging the knowledge derived from medical training and theory, rather than that potentially rooted in patient experience, which effectively limits epistemic authority to healthcare practitioners.

Without access to the same medical knowledge and adequate communication about the likely outcome of treatment decisions, families were unable to fully comprehend the gravity of what was happening, and were dependent on the trust they placed in medical professionals.

5 Police Investigations, Identity Politics, and Persistent Asymmetries in Epistemic Resources

Box Example 4

Between 1998 and 2010, the police conducted three investigations into care at Gosport War Memorial Hospital. The first investigation was initiated by Mrs. Mackenzie (daughter of Mrs. Richards) on September 27, 1998, who alleged that her mother had been unlawfully killed, and the cause of death had been wrongly recorded on the death certificate. A hospital investigation had taken place, and the hospital had admitted some liability, but Mrs. Mackenzie was concerned about a cover-up. Recording this allegation, Detective Constable (DC) Bettesworth noted that the case appeared to fall short of unlawful killing, might more properly be understood as negligence and therefore within the GMC's remit. He further noted that "we only have her word with regards to the ins and outs of the incidents and what she claims to be in this report". There was no accompanying evidence or rationale for the officer's views and Mrs. Mackenzie later stated that the police officer was "dismissive" of her and gave her the impression that he regarded her as emotional.

On October 2, 1998, DC Maddison, the investigating officer, met with the sisters Mrs. Mackenzie and Mrs. Lack. He also advised that it was not a police matter, and they should contact the GMC as the central issue was the appropriateness of the treatment. Despite forming this conclusion, the investigating officer had not: (1) identified any potential witnesses; (2) taken any statements; (3) secured medical records or the hospital investigation report; (4) undertaken a scene visit; (5) considered any forensic evidence; or (6) considered exploring other patient records for similar issues. Thus, without taking any of these actions, the first evidence file was passed to the Crown Prosecution Service (CPS) with a note describing the accusation as "death being caused by a doctor failing to give liquid by way of a drip" and stated: "my enquiries have found that this is not a necessary procedure. The doctor fully explained the procedure of placing a syringe-driver and the eventual outcome".

DC Maddison wrote a report which was passed between several senior police officers for consideration, with a number of cover notes being attached in the process. In the first DC Maddison wrote, "I have no idea why these two sisters are so out to stir up trouble", and suggested that it might be because they felt "uncomfortable about not nursing their mother". In the second, Detective Sergeant (DS) Oliver wrote that there was no case for further investigation. In the third, Detective Inspector (DI) Morgan acknowledged a "seemingly atrocious lack of care in relation to the deceased" but concurred with DS Oliver. Finally, Detective Superintendent (DSupt) Lane advised that the CPS could only make that decision once expert medical opinion had been obtained. Thus, DC Maddison sought a statement from a Macmillan nurse (a specialist nurse in palliative care) on the general use of syringe-drivers in palliative care.

On November 24, the CPS advised that there was insufficient evidence for a prosecution, that the Macmillan nurse's statement was insufficient and "properly qualified medical evidence" was needed. DC Maddison then sought such evidence from Gosport War Memorial Hospital, not recognizing that the police were then reliant on the hospital and its consultant, both potential defendants, to provide critical evidence in the case, and allowed the hospital to seek legal advice on the statement before it was submitted.

In her statement, Dr. Lord (Consultant Geriatrician and responsible for overseeing Dr. Barton) acknowledged that Mrs. Richards received continuous administration of diamorphine, haloperidol, and midazolam via a syringe-driver. This analgesia and sedation was considered necessary to address Mrs. Richards' pain, anxiety, and agitation. Mrs. Richards was provided with palliative care and, Dr. Lord said, in this context, fluids were often not used as they do not significantly alter the outcome. She further asserted that Mrs. Richards' daughters did not raise concerns about this at the time and in fact agreed to the course of action.

This statement was provided to the police on January 19, 1999, and forwarded to DI Morgan, with a cover note in which DC Maddison reiterated the view that no further investigation was warranted. On February 2, DI Morgan passed the file back to the CPS for consideration. The cover note from DI Morgan stated:

> Further evidence has been obtained ... from Dr. A Lord, a Consultant Geriatrician. He [sic] is an independent in relation to this matter in that he had nothing to do with Mrs. Richards' treatment. He is eminently well qualified to give an expert opinion in this case ... Again it would appear that this decision was taken for sound medical reasons as opposed to any wish on the Doctors part commit euthanasia in relation to Mrs. Richards.

DI Morgan's note ends by stating that Mrs. Richards' daughters were consulted about the decision to omit intravenous fluids and there was no evidence to suggest negligence and enable a successful prosecution for manslaughter. On March 11, the CPS responded: "In light of all the material provided, I do not consider there is evidence to justify a prosecution of the medical staff involved in the care of Mrs. Richards for manslaughter, or any other criminal offence". On March 17, DI Morgan informed Mrs. Mackenzie of the CPS' decision not to prosecute any of the medical staff involved in the care of Mrs. Richards and the investigation was closed.

From Mrs. Mackenzie's first contact with the police, the officer recording the allegation determined that it was negligence not unlawful killing and therefore not a police concern. The only evidence informing this judgment was Mrs. Mackenzie's testimony and the police officer was quick to cast doubt on its veracity: "we only have her word with regards to the ins and outs of the incidents and what she claims to be in this report". As outlined above, Fricker contends that assessing knowledge claims normally involves some assessment of the speaker's credibility based on their social type. The dangers of epistemic injustice are obvious; generalizations in the form of stereotypes perpetuate prejudice and discrimination. This is problematic enough in everyday life, but where assessing the veracity of knowledge claims is a professional practice—as for the police—some conscious challenge to, or at least recognition of, assumptions and the generalizations they are based on is needed. Indeed, Mrs. Mackenzie's statement about being dismissed and regarded as emotional suggests that the prejudicial stereotypes of women as over-emotional and illogical were at play.

Next, when the sisters, Mrs. Mackenzie and Mrs. Lack, met DC Maddison, the investigating officer, a similar process occurred—they were advised it was not a police matter and to contact the GMC. Again, this assessment was based entirely on the sisters' testimony as there had been no attempt to investigate the claims by this stage. Nevertheless, a file was passed to the CPS with a cover note suggesting that intravenous fluids are not always necessary when using a syringe-driver. This indicates that the same asymmetry in hermeneutical resources that was at work when informing families about treatment decisions, was also shaping the police investigation.

The cover notes that accompanied DC Maddison's report shed further light on the identity politics at play. His comment, "I have no idea why these two sisters are so out to stir up trouble", explicitly questions the daughters' motives with little attempt to verify their claims, and his suggestion that they felt "uncomfortable about not nursing their mother" invokes spurious and gendered assumptions about the proper role of a daughter in these circumstances. Against the oft-cited view that there was no case to answer, DSupt Lane highlighted the need for expert medical opinion which DC Maddison obtained from a Macmillan nurse—again signaling how the principles of palliative care framed the perceived appropriateness of treatment.

Obtaining further expert opinion from a consultant within the Trust and responsible for line managing the doctor whose practice was being questioned, significantly weights the structures of knowledge production against those making the complaint. The opinion was unlikely to be critical of Dr. Barton's practice and is in no way independent, but it also allowed the Trust to engage legal advice to limit the possibility of legal challenge.

With Dr. Lord's statement, the case was referred back to the CPS with a telling cover note. That Dr. Lord is described as "he" is not incidental but reflective of prevailing gendered assumptions about who is likely to hold authoritative knowledge on such matters. Dr. Lord is further misrepresented as being independent, which, as outlined above, she clearly was not. The cover note indicates the dominant mindset of the police that there were "sound medical reasons" for treatment rather than an attempt to euthanize Mrs. Richards.

While there were evidently prejudicial stereotypes working through this police investigation, Anderson (2012, p. 169) argues that to identify structural forms of testimonial injustice, we should move beyond the prejudice model and "consider other ways in which disadvantaged social groups can be unjustly denied credibility". One such way is through a "shared reality bias", where the perspectives of individuals who interact frequently converge (Anderson, 2012). Shared reality bias is a mundane feature of working together that helps people coordinate their actions and facilitates cooperation; however, Anderson (2012, p. 170) argues that "when groups of inquirers are segregated along the same lines that define group inequalities, the shared reality bias will tend to insulate members of advantaged groups from the perspectives of the systematically disadvantaged". Here, the same asymmetry in hermeneutical resources that denied relatives the opportunity to challenge the prescribed care, is used to justify and legitimate it. This informed the police's shared reality that no crime had been committed, a predisposition that shaped their investigation, interpretation of evidence, and ultimately steered the CPS' view that there was insufficient evidence of manslaughter.

6 Discussion

We have drawn upon The Report of the Gosport Independent Panel and applied concepts related to epistemic injustice to explore how safety concerns were heard, evaluated, and undermined. The examples demonstrate how epistemic injustice—both agential and structural—contributed to what remains an ongoing scandal: the voices of nurses were systematically muted, and both nurses and families of patients were branded as troublemakers by healthcare managers and the police.[1]

[1] In the second and third police investigations more cases were considered and more evidence collected as they attempted to address the shortcomings of the first investigation. On

6.1 Concerns of Nurses

A group of night nurses first recognized inappropriate prescribing of opioids in 1991 and repeatedly raised concerns. However, sources of agential and structural epistemic injustice were critical in perpetuating harmful practices. The nurses' credibility as knowers was continually undermined as a pervasive stereotype based on their low rank in nursing and medical hierarchies—and thus lack of knowledge—led senior staff to question whether their understanding of patients' clinical conditions was sufficient. When concerns persisted, staff leveraged their seniority to effectively silence the nurses. While these epistemic injustices were frequently enacted by senior individuals, structural injustices contributed to the systematic suppression of nurses' concerns. The healthcare system at Gosport was deeply hierarchical. Physicians' expertise was disproportionately privileged, and without any effective clinical governance systems, little evidence beyond the discredited concerns of nurses suggested systemic problems. The absence of independent oversight mechanisms also perpetuated issues. After the very clinicians they accused of malpractice had rejected these concerns, the nurses had no reporting systems through which to "blow the whistle". The cumulative effect of these rigid hierarchical systems, stereotyping and discreditation of nurses' voices, and lack of reporting mechanisms, was to foster a culture of deference toward senior clinicians, allowing inappropriate prescribing to continue unchallenged.

However, systemic hierarchical issues at Gosport, which in hindsight appear evident, may have been less obvious to doctors and management at the time. In healthcare systems, there has long been tension between the desire to create flatter hierarchies and the pressures that encourage a return to traditional power structures. A key factor in this dynamic is the way that

both occasions the CPS reached the conclusion that there was insufficient evidence to support a charge of gross negligence manslaughter. Questions about the adequacy of expert medical opinion were at the center of these decisions. From a perspective of epistemic exclusions, these investigations warrant detailed consideration as they are not easily explained by the prejudicial stereotypes that produce epistemic injustice. Instead, the police investigations and the conclusions reached by the CPS prompt more structural epistemological questions about the level of certainty and agreement clinical judgment can reach based only upon clinical records, and the standard of proof that operates within criminal law. However, we do not have the necessary space for such an exploration in this chapter and contain our focus to those events that more directly relate to the process of raising concerns and trying to have them heard and taken seriously.

senior clinicians and hospital management respond to signals of danger, such as those highlighted through complaints and incident reporting. After initial attempts to hear concerns, attitudes quickly reverted to those of a traditional hierarchy and nurses were viewed as trouble-makers. This reflects Waring's (2005) observations of the UK healthcare system at the turn of the millennia, where senior clinicians and management often saw incident reporting as a tool for shifting blame rather than a mechanism for improving safety. Waring identified how the culture of medicine, with its rejection of bureaucracy and suspicion of managerial scrutiny, works against learning from incident reporting (Waring, 2005).

In contemporary healthcare systems, lessons from the Gosport War Memorial Hospital remain relevant. Advances in practice may have lessened some power imbalances, but residual cultural norms, systemic barriers, and implicit biases can still perpetuate epistemic injustice. Flattened hierarchies do not automatically eliminate implicit biases held against less authoritative voices and agents should be wary of new and emerging prejudices as healthcare delivery evolves. One topical example from the UK National Health Service (NHS) is the relationship between doctors and physician associates (PAs). While the two professions worked in harmony for years, perceived professional "creep" has recently prompted vocal concerns from junior doctors about the safety of PAs' practice and the "toxic" debate has fractured professional relationships to the extent that the UK government has ordered a review into their practice (Triggle, 2024). As Higgins and Reitz state, "Workplaces are riddled with power and hierarchy that cannot be disappeared" (Higgins & Reitz, 2019, p. 454). Thus, the manifestations of power relations may change over time, but analyzing their impact will remain central to any claims made about safety.

6.2 *Concerns of Patients and Families*

The chapter also underscores the impact of agential and structural injustices on patients and families. Medical authorities frequently dismissed family concerns, and families struggled to gain clarity about their relatives' treatment, often being excluded from decision making processes. On a structural level, dominance of the medical profession, insufficient oversight and clinical governance, and ineffective reporting mechanisms contributed to a healthcare culture where harmful practices were allowed to continue unchecked. Carel and Kidd (2018, p. 336) highlight how "asymmetries, dependencies, and power relations can increase the vulnerability of patients to epistemic injustice" and so the hierarchical nature of medical authority, combined with institutional resistance to criticism, perpetuated a system that did not protect patients or respond to families concerns.

Our analysis shows how the root cause of patient disempowerment—the epistemic privilege enjoyed by the practitioners and institutions of contemporary healthcare services (Carel & Kidd, 2014) and relative undervaluing of patient and family experience and knowledge—is enacted through communication with patients and families and the asymmetrical use of hermeneutical resources. Carel and Kidd (2014) explain that epistemic privilege is conferred on healthcare professionals by virtue of their training and expertise, and that healthcare organizations implicitly privilege certain styles of articulating and evidencing testimonies in ways that marginalize ill persons. Nevertheless, while we acknowledge how norms of practice constrain patterns of communication, it is also potentially an area where agents have some scope to shift the ethical landscape. Attending to asymmetries in hermeneutical resources, clinicians can lessen epistemic injustice by being more explicit about the principles guiding practice and opening up space for patients and relatives to challenge the appropriateness of these. And while it is vital to hear patients' and families' perspectives and learn from previous failures, this must be done without overburdening those previously marginalized. "Epistemic exploitation" occurs when individuals with lived experiences of marginalization are unfairly expected to explain, justify, or educate others about their experiences. Typically, this benefits the person seeking education while burdening those delivering teaching because they must relive difficult experiences (Berenstain, 2016). In healthcare education and research, there has been growing enthusiasm for learning from service-users' lived experiences, but systems for properly recompensing individuals for their time and emotional labor have not kept pace, although this is now beginning to change. While payment does not reduce the pain of reliving difficult experiences, it is at least tangible recognition of the value of sharing such experiences. But once experiences have been shared, there is a responsibility on healthcare researchers, educators, and learners to take those lessons on board so that difficult experiences do not have to be repeatedly relived.

Again, these issues take on fresh significance in light of current events in the UK—Members of Parliament have been debating a proposed change in the law to legalize assisted dying and have voted in favor of the bill (Morton, 2024). Our analysis shows the need for awareness as to how implicit biases may work through this issue and that a similar shift is needed in relation to patients and families as that argued for by decolonial scholars; instead of ever more "inclusion" of marginalized others—an approach that leaves underlying structures unchallenged—we need systems that recenter the perspectives of oppressed and marginalized populations (Bhambra et al., 2018). Given the UK is at a turning point in the legalization of assisted dying (British Geriatric Society, 2024), the need to address structural barriers and individual ethical failures in communicating issues of health, palliative care, and death is felt acutely.

7 Conclusion

What does it take to settle a scandal? Epistemic justice plays a large part in this. We have shown how nurses and families were disbelieved and branded as troublemakers. This is a familiar trope applied to whistleblowers (Van Akkeren & Tarr, 2015). Processes of identifying harms have come to rely too heavily on whistleblowers—a high-stakes activity for individuals who then are systematically wronged in the process of bringing their concerns to light. Ultimately, we cannot help but agree with Higgins and Reitz (2019, p. 453): "If whistleblowing is the answer, ask a better question". Epistemic justice is central, but it is not the entire story. The inquiry fully vindicated nurses', patients', and families' concerns, yet the scandal continues to be investigated. Perhaps, following Higgins and Reitz, we have asked the wrong question, and the emphasis should be on how to prevent scandals from arising in the first instance. Here, epistemic justice would go a long way; if healthcare professionals were mindful of epistemic justice in decision making and communications with patients and families, and if healthcare executive boards, directors, and managers would embed principles of epistemic justice in organizational procedures—particularly those of complaint investigations—this might be a way to avoid at least some complaints from escalating into high-profile scandals. Healthcare professionals have an ethical responsibility to learn from inquiry reports. Inquiries crystallize contemporary societal expectations and vividly illustrate infringements of them. Given professional norms and ethical behaviors are significantly shaped by societal expectations, high-profile scandals such as these can inform what is seen as ethical behavior within the profession.

References

Anderson, E. (2012). Epistemic justice as a virtue of social institutions. *Social Epistemology, 26*(2). https://doi.org/10.1080/02691728.2011.652211

Berenstain, N. (2016). Epistemic exploitation. Ergo, *3. Article 22.* https://doi.org/10.3998/ergo.12405314.0003.022.

Berenstain, N. (2020). White feminist gaslighting. *Hypatia, 35*(4). https://doi.org/10.1017/hyp.2020.31

Bhambra, G., Gebrial, D., & Nişancıoğlu, K. (Eds.). (2018). *Decolonising the university.* Pluto Press.

Boudes, T., & Laroche, H. (2009). Taking off the heat: Narrative Sensemaking in post-crisis inquiry reports. *Organization Studies, 30*(4), 377–396.

British Geriatric Society. (2024). *Assisted dying—Where are we now and what happens next?* https://www.bgs.org.uk/blog/assisted-dying%2D%2Dwhere-are-we-now-and-what-happens-next

Carel, H., & Kidd, I. J. (2014). Epistemic injustice in healthcare: A philosophical analysis. *Medicine, Health Care and Philosophy, 17*(4), 529–540. https://doi.org/10.1007/s11019-014-9560-2

Carel, H., & Kidd, I. J. (2018). Epistemic injustice in medicine and healthcare. In I. J. Kidd, J. Medina, & G. Pohlhaus (Eds.), *The Routledge handbook of epistemic injustice* (99, 336–349). Routledge. https://doi.org/10.4324/9781315212043-33.

Dotson, K. (2012). A cautionary tale: On limiting epistemic oppression. *Frontiers: A Journal of Women Studies, 33*(1), 24–47. https://doi.org/10.1353/fro.2012.a472779

Fricker, M. (2007). *Epistemic injustice: Power and the ethics of knowing.* Oxford University Press. https://doi.org/10.1093/acprof:oso/9780198237907.001.0001

Goodwin, D. (2018). Cultures of caring: Healthcare 'scandals', inquiries, and the remaking of accountabilities. *Social Studies of Science, 48*(1), 101–124. https://doi.org/10.1177/0306312717751051

Higgins, J., & Reitz, M. (2019). If whistleblowing is the answer, ask a better question. *Journal of the Royal Society of Medicine, 112*(11), 453–456. https://doi.org/10.1177/0141076819864522

Jones, J. (2018). *Gosport War Memorial Hospital: The Report of the Gosport Independent Panel.* https://www.gosportpanel.independent.gov.uk/panel-report/.

Kent Police. (2025). *Operation Magenta.* https://www.kent.police.uk/police-forces/kent-police/areas/kent-police/about-us/about-us/op-magenta/latest-updates.

Morton, B. (2024). MPs back proposals to legalise assisted dying. *BBC News.* https://www.bbc.co.uk/news/articles/ckgzkp79npgo

Triggle, N. (2024). Streeting orders review to end 'toxic' NHS staff row. *BBC News*. https://www.bbc.co.uk/news/articles/c8rl4r36y6eo

Van Akkeren, J., & Tarr, J. (2015). Troublemakers and traitors—It's no fun being a whistleblower. *The Conversation*. https://theconversation.com/troublemakers-and-traitors-its-no-fun-being-a-whistleblower-50755

Waring, J. J. (2005). Beyond blame: Cultural barriers to medical incident reporting. *Social Science and Medicine, 60*(9), 1927–1935. https://doi.org/10.1016/j.socscimed.2004.08.055

CHAPTER 8

Divergent Attitudes Toward Ethics and Collective Responsibility in Canadian Engineering Discourse

Kari Zacharias and Robyn Mae Paul

1 Introduction

Since 2022, Canadian engineers have been reckoning with a significant piece of their culture and history: the iron ring and its accompanying ceremony, the Ritual of the Calling of an Engineer. The iron ring is a physical artifact presented in a 100-year-old ceremony to those entering the engineering profession in Canada. Engineers wear the ring on the pinky finger of their working hand, to symbolize the "obligation" made during the ceremony to honor the engineering profession and to be mindful of one's ethical responsibilities (Petroski, 1995). The ritual, written in the 1920s by author Rudyard Kipling, is administered by a volunteer organization called the Corporation of the Seven Wardens and by additional "wardens"

K. Zacharias (✉)
University of Manitoba, Winnipeg, MB, Canada
e-mail: Kari.Zacharias@umanitoba.ca

R. M. Paul
University of Calgary, Calgary, AB, Canada
e-mail: robyn.paul@ucalgary.ca

J. Hayes, S. Maslen (eds.), *Ethics in the doing of technoscience*,
https://doi.org/10.1007/978-3-032-20002-0_8

at each of the 28 "camps" throughout the country (The Corporation of the Seven Wardens, n.d.). Ceremonies are held annually at every Canadian postsecondary institution with an engineering program.

The Ritual of the Calling of an Engineer remained largely unchanged during its first hundred years. Kipling's original 1925 text was used in all camps' ceremonies with only minor alterations. The ceremony retained an air of secrecy, due to the fact that most camps restricted attendance to those who already wore iron rings, and a Freemason-esque atmosphere that involved striking an anvil with a hammer and reciting the obligation while grasping a metal chain (Home-Douglas, 2019). In 2022, the Corporation of the Seven Wardens formed a Ritual Review Committee to evaluate a potential need to change the ceremony (The Corporation of the Seven Wardens, 2022), and in 2024 the committee announced the ritual would be revised in time for its centennial the following year. A substantially revised ritual was performed for the first time in spring 2025 at camps across Canada (The Corporation of the Seven Wardens, 2025).

Professional rituals serve important roles in creating and maintaining social order and establishing collective professional understandings, even when the ritual is not tied to regulation (Song, 2018; McLean et al., 2023; Arnold et al., 2020). Wearing the iron ring and participating in the ceremony are optional activities that are not required to obtain an engineering degree, nor to become a licensed professional engineer in Canada. However, as a cultural symbol in Canadian engineering, the iron ring and the ritual are contributors to many engineers' understandings of engineering ethics and feelings of belonging in the profession (Dryburgh, 1999; Friesen, 2016; Paul et al., 2023).

This chapter uses the context of the iron ring ceremony and its change process to examine how engineers navigate ethical questions that focus on their profession as a whole. The idea that the ritual could be due for an update sparked debate among the Canadian engineering community. Throughout this discussion, members of the community expressed different interpretations of the issues at stake and the consequences of changing the ceremony. These interpretations represent divergent views of the engineering profession, engineering culture, and engineers' individual and collective ethical responsibilities. Through an analysis of public online discourse, we explore how members of the Canadian engineering community define and navigate ethical issues that concern their profession. We argue that engineers' reactions to public questioning of the iron ring ritual reveal not only how they express or mute ethical issues in their profession, but also how they perceive the bounds of their ethical responsibilities.

2 Background

2.1 *The Iron Ring and Ritual as Ethical Guides*

The iron ring and ritual serve as representations of Canadian engineering and as a moral compass for many engineers. The ring is sometimes used directly as a symbol of the engineering profession; for example, *The Iron Ring Podcast* is a show created by women engineering students at Concordia University (The Iron Ring Podcast, n.d.) and *Lady with the Iron Ring* (Lea, 2019) is a memoir. In many instances, engineers refer to the ring and/or ritual as a reminder of their ethical responsibilities. In Dan Levert's book about the 1907 Quebec Bridge collapse (Levert, 2020), he devotes the first chapters to the history of the ritual, making explicit connections between the "hubris" of the engineers involved in the disaster and the ethical lessons of the ritual and obligation. While the obligation is not a legally binding code, both the ceremony itself and the ring as an object act as ethical guides and signifiers of engineering identity.

We find parallels in other professions' ethical rituals, such as the white coat ceremony in medicine which reinforces ethical understandings while simultaneously establishing community (Huber, 2003). Even less obviously ritualistic experiences in medicine such as morbidity and mortality conferences (M&MCs) serve the manifest function (Merton, 1957) of contributing to improved patient care and also the latent functions of reinforcing group identity and sense of community (Song, 2018). In contrast to M&MCs, the iron ring ceremony is more clearly ritualistic and has a more explicit focus on group identity, as the ritual formally welcomes new engineers to the profession and reinforces a sense of solidarity through reminding them of their fallibility (Dryburgh, 1999). Indeed, establishing community among engineers is a manifest function of the ritual. The ritual's founders specifically intended to create a ceremony that would manifest a "tribal spirit" among the members of a rapidly evolving profession (Levert, 2020) and advance public recognition of engineering through establishing a shared understanding of engineers' roles and responsibilities (Milard, 1988).

While the iron ring and ritual are an ethical touchstone for many engineers, they provide general principles rather than specific guidance. The text of Kipling's original obligation centers technical and material errors, as engineers who take the oath promise to avoid "bad workmanship" and "faulty material" (The Ritual of the Calling of an Engineer, n.d.). The obligation also highlights engineers' responsibility to their profession, referencing an engineer's duty to "honourably guard" their professional reputation and "strive against" jealousy (The Ritual of the Calling of an Engineer, n.d.). As we have discussed elsewhere (Paul et al., 2023), the original ritual's focus on material safety and professional reputation obscures other aspects of ethical engineering practice, including broader sociotechnical considerations and care for the environment. More pertinently for the current discussion, the obligation's text is thematic and high-level and the ethical guidance provided by the obligation does not indicate specific courses of action.

The ritual also leaves open questions about engineers' individual and collective ethical responsibilities. Traditional engineering ethics research and teaching has focused on engineers as individual moral agents, examining ethical responsibility and decision making through individual processes and actions that can be undertaken at a "micro" level (Herkert, 2005; Riley, 2008). However, engineering ethics includes multiple overlapping dimensions of individual and collective responsibility (Davis, 2006). Engineers are part of many types of collectives, from the teams and organizations in which they work, to their profession, to broader societies and ecosystems (Zacharias et al., 2025). The ritual encourages a strong collective sense of fellowship and community, while simultaneously focusing its ethical direction on individual actions. Thus, while the ritual and ring are important touchstones for many engineers, they are incomplete guides to ethical decision making and action and may not equip engineers to engage in broader ethical discussions.

2.2 Roles of Ethical Codes and Guidelines in Engineering Culture

Professional engineering identity tends to be grounded in analytical problem-solving, logical thinking, and strong technical skills (Faulkner, 2009; Macdonald-Roach et al., 2024; Tonso, 2014). This professional identity is built through math and physics foundations that are often viewed as predictable, logical, and objective, and exist in a dualistic tension that excludes social skills and subjective considerations (Faulkner, 2000, 2015). Furthermore, engineers' gender, race, and other aspects of their personal identity influence the ease with which they can take on the professional identity of an engineer (Hatmaker, 2013; Cardador, 2017). Wendy Faulkner's work has consistently found that men are more easily able to shape organizational cultures, draw boundaries of engineering technical expertise, and be seen as "real" engineers who "belong"

(Faulkner, 2009, 2015). This pattern continues today: although racial and gender diversity has increased in the profession, racialized people and white women feel a disproportionately lower sense of belonging in engineering professional settings, and racialized women experience discrimination and barriers to licensing (Radebe et al., 2023).

Embedded within engineering cultures is an often unstated belief that engineering exists as an apolitical, neutral space, where social, political, and cultural considerations taint engineering work (Cech & Waidzunas, 2011; Cech, 2014). This dominant expression of engineering culture also portrays ethical decision making as purely rational, without the influence of emotions (Lönngren et al., 2020). Yet, emotions are essential to sensitizing engineers to risk, morals, and ethical considerations of technologies (Roeser, 2012). Engineers who discard emotions as irrational during problem-solving risk being unable to consider others' perspectives or imagine possible ethical impacts of designs. Accordingly, cultures of engineering that emphasize technocratic approaches over more sociotechnical approaches rarely provide sufficient support for engineers to navigate ethical decisions regarding risk, uncertainty, and disclosure of information (Kim et al., 2021).

Engineers often resolve this struggle by relying on codes of ethics. However, this process, as well as the ethical codes themselves, is insufficient "to guide novice engineers through ethically ambiguous situations" (Rottmann et al., 2018, p. 1). Codes and other formalized ethical guidelines strongly influence engineers' attitudes and work, despite their limitations (Seniuk Cicek et al., 2025). Although engineering codes of ethics can be ill-defined (Slaton, 2012), they create and reinforce boundaries between those who can practice "proper technical conduct" (Riley, Slaton, & Herkert, 2015a, p. 4) and those who cannot. Such boundaries also reflect dominant engineering narratives that define positivistic, seemingly objective approaches as the most appropriate ways of solving engineering ethical problems (see Paul et al., 2023, for more discussion on ethical boundaries created by the iron ring ceremony). Thus, while codes are imperfect guidelines for engineering work, they can nevertheless empower engineers to approach ethical decision making with a strong belief in their own authority and ability to make "objective" ethical choices (Downey, 2012; Bauschpies et al., 2018). As Riley and Lambrinidou (2015) argue, the emphasis on "holding the public paramount" or "protecting the public" in many engineering codes of ethics is rendered meaningless when engineers undervalue or ignore communities' "perspectives and knowledges that diverge from dominant engineering points of view" (p. 2).

2.3 *Approaches to Navigating Ethical Dilemmas*

In the absence of other clear and actionable ethical guidelines, many engineers rely on organizational policies and practices (Rottman et al., 2021). However, similar to the codes discussed above, reliance on existing practices can lead to limiting or inflexible outcomes (Riley, Slaton, & Herkert, 2015a) and can be an additional challenge when the root of an ethical dilemma comes from pressures within the organization itself (time, cost, efficiency, etc.) (Adams, 2020). Engineers may also lose faith in their organizational structures, especially if they come to regard regulators as "just bureaucrats" who are "indistinguishable from politicians and more than willing to harm the public for personal gain" (MacDonald-Roach et al., 2024, p. 18).

Alternative ways engineers have navigated ethical dilemmas include stories and connections with peers. For example, an "intertwining of abstract and narrative forms of reasoning" during a discussion allowed engineers to debate implications of engineering disasters for professional practice (Maslen & Hayes, 2020, p. 208). Additionally, women and racialized engineers will leverage their networks when they encounter ethical dilemmas, reaching out to family, friends, and colleagues for social and emotional support (Rottmann et al., 2021). This highlights the importance of community-based professional networks in providing support for engineers as they navigate ethical dilemmas in the workplace. By leveraging and supporting these networks, "we may move closer to achieving the socially just, but as yet symbolic promise inherent in the public paramountcy clause of engineers' ethical codes" (Rottmann et al., 2021).

Overall, engineering educational and professional cultures promote strong values of objectivity and technocratic solutions, leaving engineers without skills to engage in solving ambiguous ethical dilemmas. The emphasis on ethical codes or obligations such as the iron ring ceremony is not always helpful, as they can be fuzzy and ill-defined, but still give engineers a sense of authority that can lead them to be dismissive of others' perspectives in ethical dilemmas. This widens the artificial gap between ethics and equity, further institutionalizing and normalizing injustices of engineering. Additionally, the emphasis on apolitical and neutral approaches to engineering pushes engineers to avoid engaging in most ethical dilemmas. To break this cycle, educational and professional engineering must go beyond the technical factors that drive engineering decisions, acknowledging the critical ethical impact of values, politics, emotions, and collective responsibility. In this chapter, we demonstrate the ways in which contributors to online discourse about the iron ring ceremony engage with ethical dilemmas, and how these discussions both perpetuate the current engineering culture and suggest new ways forward.

3 Methods

In this chapter, we use the context of the iron ring and surrounding controversies about the ritual to explore how engineers interpret and navigate ethical dilemmas that relate to their profession as a whole. We ask, how does the iron ring discourse illuminate how members of the engineering community express or mute ethical questions related to the future of the engineering profession in Canada? To address this question, we analyzed publicly available online discussions about the iron ring ceremony, including proposed changes to the ritual. This is in line with research which demonstrates there is a strong relationship between professional discourse and social practice (Irimiea, 2017). We did not verify the identity or status (student, professional engineer, etc.) of any contributors, but our searches targeted sources that were written by self-identified engineers in Canada, or aimed at engineering audiences. Although our data might include non-engineers, professional discourse is defined to broadly include intraprofessional (within a specific profession), interprofessional (across two professions such as engineering and medicine), and lay-discourse including non-professionals (Kong, 2014 as cited in Irimiea, 2017). All data was taken from comments, discussions, posts, and articles that were publicly available with no expectation of privacy from the contributors, and therefore appropriate for research use in accordance with Canadian research ethics guidelines (TCPS 2, 2022). The table in the appendix summarizes data sources and the date of publication.

We identified articles, discussion forums, and posts through a series of searches conducted in August 2024 on Google and LinkedIn. In our Google searches, we used the search terms "engineering" + "iron" + "ring" + "ceremony" and reviewed the results page by page until they lost relevance. We analyzed both the text of the article or post and any comments that were attached to it through a qualitative content analysis process (Elo & Kyngäs, 2008). To further target discussions among the professional engineering community, we searched within LinkedIn, which is the primary social media platform used in professional contexts in Canada, to explore discussions in existing informal professional support networks (Rottmann et al., 2021). Our LinkedIn searches used the platform's search tool and the search term "iron ring." While most sources engaged specifically with the Wardens' Ritual Review Committee, established in 2022, other sources predated this. We included these older sources in our analysis to demonstrate that the ethical questions surrounding the ring and ceremony arose well before the current moment.

After gathering the data, we first explored the discourse through rough, open coding. We reviewed the data individually, noting themes, arguments, and striking language. Following this first pass, we identified contributors' different interpretations of the nature and importance of the debate around the iron ring as a central phenomenon of interest. Our subsequent analysis focused on what ethical dilemmas, if any, were understood or expressed by contributors. This expression was not always explicit or positive: in many cases, contributors communicated their understanding of a dilemma by indirect implication, or by denying that it existed. Contributor quotes where there was discussion in ethical dilemmas were coded, with 14 initial categories emerging from the data, which were grouped into four themes. Quotes in the findings section are cited based on our data identifier and the source, for example [8, EngTips] refers to source #8 (see Table in Appendix) which is the EngTips forum.

4 Findings

The overarching dilemma that engineers navigated in this case may appear to be a single question, e.g., "what should we do about the iron ring ceremony?" However, contributors to the discourse disagreed not only about the answers to this question but also about the underlying ethical questions themselves. Contributors interpreted the question of the iron ring as a concern about sexism, colonialism, racism, religiosity, transparency, accessibility, humility, and the nature of engineering ethics. Here, we have organized and described three fundamental framings.

4.1 *"A Reflection of His Time and Place": Questioning the Value of Tradition*

Many contributors framed the question of what to do about the iron ring ceremony as a debate about the value of tradition within engineering. In these cases, the central issue was *how engineering's past informs—or ought to inform—its present and future.* This framing often appeared when contributors addressed colonialism in the iron ring ceremony: either due to the association of Rudyard Kipling (the ritual's author) with British imperialism and white supremacy (Parry, 1996) or due to the Masonic atmosphere and symbolism of the ritual itself (Langins, 2019), including the prominent use of a chain as a ritual artifact. Contributors asked what ethical responsibility current-day engineers have to the founders of their profession and their traditions; what responsibility they have to address past harms; and to what extent engineers should or must embrace change as an ethical obligation.

Many contributors emphasized the importance of preserving tradition, often alluding to the potential dangers of overturning long-standing practices. Changing the ceremony would leave engineers with "nothing to hang [their] hat on" [1, ESE], needing to continuously "beg for future changes" in the coming years [8, EngTips]. More specific arguments warned that changing the ritual would lead to its demise, directly refuting Wardens' stated intention of reestablishing the ceremony to last in "a modern Canadian society" (The Corporation of the Seven Wardens, 2022). One commenter composed their own poem to this effect, casting those arguing for change as "blinded by their own conceit" [8, EngTips].

Contributors sometimes disagreed over whether issues could or should be consigned to the past, as opposed to viewing them as contemporary concerns for the engineering profession. This was particularly apparent in discussions of colonialism. In one public forum, commenters dismissed arguments for changing the ceremony due to its colonial connotations, arguing that the only place for colonialism "a hundred plus years past [its conclusion]" was in "a purely historical discussion" [2, Reddit], or suggesting that the ceremony is "a reflection of [Kipling's] time and place" [2, Reddit]. Others recognized Kipling himself as imperialist or racist, but did not see this as a reason for change. As one commenter put it bluntly, "Yeah, [Kipling] sucks, but that seems to be for reasons largely divorced from the ring ceremony" [2, Reddit]. Some contributions in the same thread pushed back against this rhetoric, including by explicitly pointing out that colonialism is "absolutely not a thing of the past" [2, Reddit]. Yet, most commenters in this forum addressed the framing of the iron ring as a debate about colonialism by dismissing or downplaying the contemporary relevance of the issue to the engineering profession.

These discussions reflect ignorance of the endemic and ongoing nature of colonialism (Tuck & Gaztambide-Fernández, 2013; Godlewska et al., 2020), as well as historical and contemporary relationships between colonialism and engineering (Headrick, 1988; Martin & Hoffman, 2011;

Perry, 2016). They also demonstrate a dominant understanding of engineering as inherently apolitical and objective: raceless, classless, genderless, and otherwise unconcerned with identity and culture (Cech, 2013; Secules, 2019; Leydens & Lucena, 2018; Seron et al., 2018). Viewing engineering as separate from social and political concerns allows commenters to draw a line between the practice and effects of colonialism and the profession of engineering.

In other contexts, contributors identified reasons to part with the tradition of the iron ring ceremony in relation to the ceremony's colonial origins or values. Rather than emphasizing the risks associated with departing from tradition, they portrayed change as an ethically necessary part of "an inclusive and progressive profession" [56, LinkedIn]. Canadian engineering has moved "far beyond this oath" [60, LinkedIn], and revision is necessary to bring the ceremony and experience in line with the profession. Notably, under this framing, the engineering profession itself was not problematized. In this sense, these comments align with the apolitical nature of those mentioned above. However, these comments do not separate engineering from sociocultural issues in the same way that those arguing for tradition do. Contributors frame progress in society and the profession as parallel and argue that the ritual must change to keep up with both. In the words of one Warden, "The way you are taught is changing ... All these things are getting all these inputs and all of the sudden, [the Ritual] isn't consistent with that" [59, PEO].

4.2 *"That's Just How It Is": Roles for Diversity, Equity, and Inclusion in the Engineering Profession*

Both advocates for change and supporters of the existing iron ring ceremony sometimes framed the discussion in terms of diversity, equity, and inclusion (DEI). Those arguing for change made connections between the iron ring ceremony as a representation of engineering ethics and equity concerns such as structural sexism and racism within the engineering profession. These contributors often cited personal experiences of discomfort or microaggressions during the ritual. Those opposed to change disputed the value of DEI initiatives, dismissing arguments about systemic inequalities in engineering and insisting on separation between engineering ethics and DEI. While these two groups were diametrically opposed in their opinions about the ceremony, they shared a common reframing of the ethical dilemma as a question about *what the place or purpose of DEI in the engineering profession ought to be.*

As discussed above, ethics and equity have traditionally been considered separately in engineering contexts (Riley, Slaton, & Herkert, 2015a; Riley, Slaton, & Pawley, 2015b), though equity-minded engineering educators and researchers have articulated important connections between them (Riley & Lambrinidou, 2015; Cech, 2013). Engineers may be more likely to perceive systemic inequities when they have a personal connection to an issue, but can also be hesitant to speak out due to fear of recrimination (Rottmann & Reeve, 2020). This was reflected in comments that dismissed criticism of the ceremony's lack of diversity or exclusionary nature on the grounds that these concerns were separate from engineering ethics. For these contributors, their moral attitudes about engineering work were rooted in their "pride, honour and responsibility to a calling" [1, ESE] that was reinforced by the iron ring ceremony. They maintained pride in their profession regardless of any feelings of unease during the ritual or "stuff about Kipling" [4, Reddit], because of "the fact that [they] have an obligation as an engineer" [56, LinkedIn]. In other words, "the symbolism of being ethical and upholding our standards won out" over potential equity concerns [4, Reddit].

These comments demonstrate a connection between framings of engineering ethics as distinct from equity and social justice concerns, and an understanding of engineering as rational and objective (Seron et al., 2018; Leydens & Lucena, 2018). Interestingly, the commenters above dismissed some feelings (e.g., unease) as irrelevant, while embracing others (e.g., pride), that aligned with the ritual's intent. Others who made a connection between equity and ethics noted that conversations immediately after the ceremony "revolved around [participants'] disappointment and disgust" instead of "their pride in their accomplishments" [15b, UBC]. One contributor specifically connected equity and ethics by acknowledging how exclusivity can prevent ethical understandings, where "if the ceremony excludes new graduates […] then our next generation of engineers may miss the important message that the ritual was created to communicate" [7, EngTips]. Some commenters directly confronted those who shared their misalignment of feelings. For example, when one commenter remembered feeling "vaguely disgusted" [60, LinkedIn], another replied saying that "disgust" was "a bit of a harsh reaction" as the ritual is "not what people might glean from" it, but rather it is about "working honourably to protect one's reputation and the reputation of the profession" [60, LinkedIn]. In contrast, some cited these negative feelings as reason for change, arguing that "if there's anyone being actively harmed by participating in the ceremony, I think that warrants changing things up" [56, LinkedIn].

While most commenters agreed that equity and diversity were positive goals for the profession to pursue, or even core values for engineering, others argued that true equity in engineering is simply not possible. These comments described proposed changes as "simply replacing one taint with another" [7, EngTips] because "every idea has pros and cons that will help some and harm others" [8, EngTips] and thus any change must "avoid attempting to satisfy every potential interest" [58, LinkedIn]. The underlying premise here is that if engineering cannot help everyone, there is no point in trying to help anyone—one commenter even saying that the ring is "non-inclusionary" to those who don't have hands to wear it on, thus aiming for diversity and inclusion in itself is unethical and will lead to "outright destruction" [7, EngTips]. While these comments predate the Trump administration's anti-DEI policies by several years, they represent similar motivations and ideals, which have emerged in a variety of contexts (Ng et al., 2025; Prasad & Śliwa, 2024). Within engineering, this anti-DEI rhetoric reflects discriminatory and exclusionary attitudes about who can be an engineer, whereby

> that which is thought or performed by [...] the gender-nonconforming person, by the person of physical or intellectual "disability", the old or the poor person, and often the person of color or the female-gendered person, is frequently illegible as rigorous engineering. (Slaton et al., 2019)

Conversations that framed the iron ring discussion as a question about DEI also demonstrated how equity-minded engineers can be put in "a difficult position" [10, The Conversation]. A newspaper article highlighted how this dilemma places a burden on new graduates as they are "confronted with" balancing the "harms" of what the ceremony represents and the symbol of "an engineer's dedication to being ethical and moral" (Hurley, 2023). Racialized engineers and white women also face additional burdens when navigating ethics and equity-based dilemmas (Rottmann & Reeve, 2020), which for some contributors made them disinclined or unable to continue pushing for change. One commenter reflected on her 1992 ceremony, where she raised an issue about the term "brotherhood," being used in the ceremony, but "was dismissed" and she lacked "the energy to pursue it further" [53, LinkedIn]. These dismissals communicated to participants that they were "not welcome to enter into the industry" [15, Ubyssey]. As one commenter put it, "to me it was a value judgement they were communicating that including me wasn't valuable to them" [4, Reddit]. Others lamented feeling stonewalled, that "there's not much that we could do, that's just how it is" [61, LinkedIn]. This continuous cycle of exclusion, dismissals, and dilemmas, led to minoritized engineers self-identifying as "battle hardened, having dealt with so much crap along the way" [8, EngTips]. Still, some contributors found hope in the possibility for change. A self-identified senior engineer who attended the ritual to present a ring to her son recalled encountering a lineup of female students also hoping she would present them with their rings. She expressed hope that changes to the ceremony would reduce the "unfairness" experienced by these students [48, LinkedIn].

4.3 "Our Moral Obligation": Individual and Collective Engineering Responsibility

A third framing of the dilemma focused on *individual and collective ethical responsibility in engineering professional practice.* This framing not only compared individual and collective duties and choices, but also questioned the meaning of collective ethical responsibility. The iron ring itself was at the center of much of this discussion, as a tension emerged between the ring as a commitment to individual action and as a symbol of the collective profession.

A few commenters called out the need to think of ethics "beyond the individual responsibility to design" safely, as engineers need to "debate whether it is enough to commit to build pipelines that won't leak [...] when the products flowing through those engineers worlds pose a fundamental threat to humankind" [39, UBC]. Specifically, they stated that the iron ring ceremony "may fail to properly emphasize [that] doing the right thing might be very difficult to do on your own" [LinkedIn, 48]. Although these commenters acknowledged the importance of understanding individual responsibilities, they urged that engineers must also know "our collective responsibility as a profession" [UBC, 39]. In contrast, other commenters feared having "every statement and decision second-guessed decades afterward" [8, EngTips] or being "judged by the moralists of 2122" [2, Reddit]. In these arguments, individualistic perspectives perpetuated a belief in value neutrality of engineering and technology arguing that engineering responsibility ends with a completed design and doesn't extend to future collective consequences (Bauschpies et al., 2018). As one contributor argued, "it's one thing to gravely assume responsibility for workmanship and quite another to accept the potential consequences brought about by that work" [10a, The Conversation].

Other comments highlighted different understandings of collective responsibility. There was consensus across many commenters that the iron ring brings a sense of belonging to engineering in Canada, where it "is an embedded part of the profession" [11, Randstad] and "is synonymous with being an engineer in Canada" [15, UBC]. Many Canadian engineers have encountered other "Ring-wearers" in different parts of the world and experienced "an instant connection [...] from the knowledge that our profession is one which has an impact on society" [34, Steemit]. Yet, this sense of community and belonging was interpreted through an emphasis on individual commitment: "The iron ring oath is shared by a large community of engineers. Our rings represent individual commitments to doing what is right" [48, LinkedIn].

Conceptions of collective responsibility sometimes focus on commitment to the public and elsewhere on obligation to one's fellow engineers. The difference between responsibility to the public and the profession most often arose through comments discussing the ceremony's secrecy and lack of transparency. Most official statements portray the ceremony as "private" (Hurley, 2023) or "modestly discrete" [13, Sheaf] as opposed to "secret." Those who believe in this discretion tended to emphasize engineers' responsibility to their peers, advocating that the ceremony is meant to be a private pledge "before other engineers ... rather than the public" [32, ASEE]. In contrast, commenters who interpreted the collective oath as a commitment to the public believed the ceremony should be more open, to "explicitly acknowledge our moral obligation to serve society" [32, ASEE].

Some commenters suggested that individuals could resolve their own ethical dilemmas around the iron ring ceremony by simply choosing not to attend. Here, commenters argued that because there "is nothing compulsory about participating in the Iron Ring Ceremony" those who morally disagreed to the ceremony could simply "opt out" [1, ESE]. Wardens who oversee the ceremony have aligned with this, responding to criticism in 2019 by suggesting that critics could simply "not...come, because [the ceremony] is optional" [15, UBC].

5 Discussion

Our analysis of this online discourse demonstrates that the iron ring and ceremony are valuable ethical guides for Canadian engineers. Regardless of their attitudes toward changing the ritual, most commenters emphasized the centrality of the iron ring to Canadian engineering identity and the importance of the obligation and ceremony in defining and reinforcing engineers' ethical responsibilities. However, contributors perceived engineering ethics in very different ways. For some, especially those who argued in favor of retaining the original ceremony, engineers had an ethical responsibility to uphold existing norms, of which the iron ring was a strong symbol. For others, particularly those who favored revising the ceremony, continuous reflection and consideration of multiple perspectives were at the core of engineers' ethical duties.

Those who favored keeping the original iron ring ceremony emphasized the collective responsibility to uphold engineering traditions and the ethical duty to take professional pride in engineering. These commenters also emphasized the iron ring should be maintained as a private oath and collective commitment before other members of the profession. These perspectives emphasize that the ethical guidelines embodied by the ritual and ring, although ill-defined, are meant to reinforce a boundary around those who are professional engineers and thus able to conduct "proper technical conduct" (Riley, Slaton, & Herkert, 2015a, p. 4), explicitly omitting perspectives from those who exist outside the boundary (Seniuk Cicek et al., 2025; Slaton, 2012). Alternatively, those who were arguing for changing the iron ring ceremony emphasized engineers' collective responsibility to acknowledge the historical harms of engineering; critique normative engineering culture to ensure the profession values equity and inclusivity; and change harmful practices. This perspective reflects an underlying belief that engineering is not neutral, and holds that by acknowledging political, cultural, and emotional considerations of the iron ring ceremony, engineers are able to critically evaluate the risks and imagine possible ethical impacts based on others' perspectives (Roeser, 2012; Cech & Waidzunas, 2011; Kim et al., 2021; Bauschpies et al., 2018). Collective responsibility is emphasized as a responsibility to understand and be aware of others' perspectives, particularly when they are in tension with the normative culture.

This disparity in interpretations of the ceremony points to the difficulty inherent in reconciling a 100-year-old ritual with contemporary understandings of engineering ethics. Kipling's original ritual text had multiple purposes, including building community among members of the engineering profession and establishing a shared set of ethical principles (Levert, 2020). Scholars have critiqued Kipling's understanding of engineering as old fashioned even by 1920s standards (Langins, 2019), and considerable changes to both the profession of engineering and engineering ethics over the past century have certainly rendered the ritual's portrayal of ethical responsibility outdated in some respects. When modern engineers who embrace sociotechnical consequences and macro-ethical approaches (Herkert, 2005), or who seek to integrate ethics and equity in engineering work (Rottmann & Reeve, 2020), overlay their ethical understandings onto the iron ring ceremony, it may be easy for them to conclude that the ritual is in need of an update. In contrast, engineers whose understanding of their own ethical obligations is more rooted in traditional framings of engineers as individual moral agents may take less of an issue with the ceremony's portrayal of engineering, and find significant value in its emphasis on building community and establishing responsibility toward the profession itself.

As a ceremony that is intended to be relevant to all engineers, the ritual's ethical guidance is necessarily general. Like many engineering codes of ethics, the iron ring ceremony is open to interpretation, and the obligation alone is insufficient to help engineers navigate ethically ambiguous situations (Rottmann et al., 2018). The discourse around potential changes to the ceremony suffers from this same issue of generality: while each of the three framings we discuss above represents a valid approach to a complex ethical situation, each is also incomplete. Some contributors were open to considering multiple framings of the discussion around the iron ring ritual, but many simplified the issue down to one of these three perspectives. This is unsurprising in the context of short-form online posts, but we found it contributed to collapsing the larger issue in ways that sometimes led to misinterpretation or avoidance of important questions.

An extreme example of this oversimplification was one commenter's stance against updating the ceremony, arguing that changes on the basis of equity and inclusion would be worthless since the ritual could never become perfectly inclusive of everyone. In addition to expressing a fundamental conservatism that has often been attributed to engineers, though rarely studied in detail (Smith et al., 2025), this comment demonstrates a diminishing and twisting of the ethical question at hand. The issue of whether the iron ring ceremony should be updated is narrowed to a question about inclusivity in engineering, before being re-expanded into a question about the feasibility of so-called DEI initiatives. The commenter then addresses the issue in a way that elides its ambiguity and proclaims the commenter's own authority and objectivity (Downey, 2012; Bauschpies et al., 2018).

We found an important counterpoint in certain types of online forums, particularly those that centered women engineers. Within our limited dataset, spaces like a Women Engineers subreddit had a greater sense of openness than spaces broadly designated for all engineers. Commenters more often asked clarifying questions, acknowledged ambiguity, and emphasized they were speaking from their own perspectives. Rottmann et al.'s (2021) research aligns with this, as they found women and racialized people tended to rely on diverse professional networks to discuss ambiguous situations, whereas white men tended to fall back on rules and policies. Their recommendation to recognize the importance of "community-based professional networks" (Rottmann et al., 2021, p. 6) is valuable in the context of the iron ring as well, as it points those involved in the change process toward potential avenues for nuanced discussion.

6 Conclusion

This chapter analyzes public online discussions about the iron ring and ritual to help us understand how members of the Canadian engineering community define and navigate ethical issues. We found that reactions to public questioning of the iron ring ritual emphasized how engineers perceive the bounds of their ethical responsibilities and differing meanings of collective responsibility. Those who favored retaining the original ceremony tended to avoid ambiguity, have strongly set boundaries, and emphasized collective responsibility as a duty to uphold existing normative tradition and culture, whereas those supportive of change tended to emphasize the ethical importance of reevaluating cultural boundaries, and emphasized a collective responsibility to consider different perspectives in ethical dilemmas. These findings parallel research showing that broad statements in engineering codes and oaths that emphasize the need for engineers to "hold paramount" the "health, safety, and welfare of the public" (APEGA, 2022) do not provide clarity for engineers when navigating ethical dilemmas (Rottmann et al., 2021; Slaton, 2012). Rather, supporting the development of sociotechnical skills which allow for nuanced, subjective, and emotional discussions will develop engineers who are able to navigate challenging ethical decisions.

These results suggest several lessons for engineers in professional practice. First, mainstream understandings of engineering ethics among professional communities are diverse and changing. The discussions we observed demonstrated varying conceptions of ethical responsibility that ranged from micro to macro and included numerous constituencies. Companies and professional organizations must recognize and navigate this diversity of ethical thought. Second, there is opportunity for those who create or manage ethical guidelines, including but not limited to the iron ring ceremony itself, to be explicit about the values that shape their work. Codes and obligations on their own will never be sufficient ethical guides for practicing engineers, but careful development may shrink the broad range of interpretations that we observed in this case. Finally, our findings emphasize the importance of encouraging engineers, especially minoritized engineers, to connect with communities of their peers. These communities offer not only support, but also spaces where engineers can collaboratively navigate ethical dilemmas. While this is particularly important for engineers who may not easily find community in their everyday work environments, facilitating strong peer connections will encourage reflection that can support all engineers.

Appendix

Table Data sources

#	*Citation*	*Details (if applicable)*	*Date of publication*
1	ESE	Environmental Science and Engineering Magazine	July 2023
2	Reddit	AskEngineers	December 2022
4	Reddit	WomenEngineers	February 2023
5	Reddit	WomenEngineers	May 2023
6	Reddit	WomenEngineers	April 2023
7	EngTips		January 2023
8	EngTips		December 2022
9	ESE	Environmental Science and Engineering Magazine	March 2023
10	The Conversation		December 2022
10a	The Conversation		April 2021
11	Randstad		February 2024

#	*Citation*	*Details (if applicable)*	*Date of publication*
12	Medium		March 2024
13	Sheaf	University of Saskatchewan newspaper	January 2013
15	Ubyssey	UBC Newspaper	February 2020
15a	UBC	UBC Engineering Undergraduate Society	March 2019
15b	UBC	Open Letter from UBC to Camp Wardens	March 2019; March 2016; July 2018
15c	UBC	UBC Engineering Undergraduate Society	February 2020
22	Peterborough Examiner		July 2024
23	Queens Journal		March 2023
26	ESE	Environmental Science and Engineering Magazine	March 2024
32	ASEE	American Society of Engineering Education	December 2019
33	LiveJournal		March 2006
34	Steemit		January 2019
39	UBC	Blogs	March 2017
47	LinkedIn		November 2023
48	LinkedIn		July 2023
51	LinkedIn		February 2024
53	LinkedIn		April 2024
56	LinkedIn		December 2022
57	LinkedIn		December 2022
58	LinkedIn		September 2022
59	PEO	Professional Engineers of Ontario	Spring 2023
60	LinkedIn		September 2022
61	LinkedIn		September 2022
62	LinkedIn		September 2022

References

Adams, T. L. (2020). 'This happens all the time': Organizations, rationalization and ethical dilemmas in engineering. *Work, Employment and Society, 34*(6), 985–1003. https://doi.org/10.1177/0950017020902968

APEGA. (2022). *Professional practice guideline: Ethical practice.* APEGA. https://www.apega.ca/docs/default-source/pdfs/standards-guidelines/ethical-practice.pdf?sfvrsn=78261e0b_14

Arnold, M. H., Komesaroff, P., & Kerridge, I. (2020). Understanding the ethical implications of the rituals of medicine. *Internal Medicine Journal, 50*(9), 1123–1131. https://doi.org/10.1111/imj.14990

Bauschpies, W., Holbrook, J. B., Douglas, E. P., Lambrinidou, Y., & Lewis, E. Y. (2018). Reimagining ethics education for peace engineering. *2018 World Engineering Education Forum—Global Engineering Deans Council (WEEF-GEDC)*, 1–4. https://doi.org/10.1109/WEEF-GEDC.2018.8629655

Cardador, M. T. (2017). Promoted up but also out? The unintended consequences of increasing women's representation in managerial roles in engineering. *Organization Science, 28*(4), 597–617. https://doi.org/10.1287/orsc.2017.1132

Cech, E. A. (2013). The (mis)framing of social justice: Why ideologies of depoliticization and meritocracy hinder engineers' ability to think about social injustices. In J. Lucena (Ed.), *Engineering education for social justice* (Vol. 10, pp. 67–84). Springer. https://doi.org/10.1007/978-94-007-6350-0_4

Cech, E. A. (2014). Culture of disengagement in engineering education? *Science, Technology, & Human Values, 39*(1), 42–72. https://doi.org/10.1177/0162243913504305

Cech, E. A., & Waidzunas, T. J. (2011). Navigating the heteronormativity of engineering: The experiences of lesbian, gay, and bisexual students. *Engineering Studies, 3*(1), 1–24. https://doi.org/10.1080/19378629.2010.545065

Davis, M. (2006). Engineering ethics, individuals, and organizations. *Science and Engineering Ethics, 12*(2), 223–231. https://doi.org/10.1007/s11948-006-0022-y

Downey, G. L. (2012). The local engineer: Normative holism in engineering formation. In S. H. Christensen, C. Mitcham, B. Li, & Y. An (Eds.), *Engineering, development and philosophy* (Vol. 11, pp. 233–251). Springer. https://doi.org/10.1007/978-94-007-5282-5_14

Dryburgh, H. (1999). Work hard, play hard: Women and professionalization in engineering-adapting to the culture. *Gender and Society, 13*(5), 664–682. https://doi.org/10.1177/089124399013005006

Elo, S., & Kyngäs, H. (2008). The qualitative content analysis process. *Journal of Advanced Nursing, 62*(1), 107–115. https://doi.org/10.1111/j.1365-2648.2007.04569.x

Faulkner, W. (2000). Dualisms, hierarchies and gender in engineering. *Social Studies of Science, 30*(5), 759–792. https://doi.org/10.1177/030631200030005005

Faulkner, W. (2009). Doing gender in engineering workplace cultures. I. Observations from the field. *Engineering Studies, 1*(1), 3–18. https://doi.org/10.1080/19378620902721322

Faulkner, W. (2015). 'Nuts and bolts and people' gender troubled engineering identities. In S. H. Christensen, C. Didier, A. Jamison, M. Meganck, C. Mitcham, & B. Newberry (Eds.), *Engineering Identities, epistemologies and values* (Vol. 21, pp. 23–40). Springer. https://doi.org/10.1007/978-3-319-16172-3_2

Friesen, M. (2016). Professional integration as a boundary crossing: Changes to identity and practice in immigrant engineers in Canada. *Engineering Studies, 8*(3), 189–211. https://doi.org/10.1080/19378629.2016.1247450

Godlewska, A. M. C., Schaefli, L. M., Forcione, M., Lamb, C., Nelson, E., & Talan, B. (2020). Canadian colonialism, ignorance and education. A study of graduating students at Queen's University. *Journal of. Pedagogy, 11*(1), 147–176. https://doi.org/10.2478/jped-2020-0008

Hatmaker, D. M. (2013). Engineering identity: Gender and professional identity negotiation among women engineers. *Gender, Work and Organization, 20*(4), 382–396. https://doi.org/10.1111/j.1468-0432.2012.00589.x

Headrick, D. R. (1988). *The tentacles of progress: Technology transfer in the age of imperialism, 1850–1940*. Oxford University Press.

Herkert, J. R. (2005). Future directions in engineering ethics research: Microethics, macroethics, and the role of professional societies. In *Engineering ethics*. Routledge. https://www-taylorfrancis-com.uml.idm.oclc.org/reader/download/b5dffc2b-3870-432c-a400-b908aaa7a297/chapter/pdf?context=ubx

Home-Douglas, P. (2019, December). Don't call it a celebration. *ASEE Prism*. http://www.asee-prism.org/dont-call-it-a-celebration/

Huber, S. J. (2003). The white coat ceremony: A contemporary medical ritual. *Journal of Medical Ethics, 29*(6), 364–366. https://doi.org/10.1136/jme.29.6.364

Hurley, J. (2023, February 17). Canadian engineers wear the iron ring with pride. Why some are sparring over the mysterious ceremony that comes with it. *Toronto Star*. https://www.thestar.com/news/gta/canadian-engineers-wear-the-iron-ring-with-pride-why-some-are-sparring-over-the-mysterious/article_bac3c9a6-a047-58a7-97b5-126691bcc511.html

Irimiea, S. B. (2017). Professional discourse as social practice. *European Journal of Interdisciplinary Studies, 3*(4), 108–119. https://revistia.com/files/articles/ejis_v3_i4_17/Silvia.pdf

Kim, D., Howland, S., & Jesiek, B. (2021). Encountering engineering ethics in the workplace: Stories from the trenches. *2021 ASEE Virtual Annual Conference Content Access Proceedings*, 37030. https://doi.org/10.18260/1-2%2D%2D37030

Kong, K. (2014). *Professional discourse*. Cambridge University Press.

Langins, J. (2019). The sons of Martha versus the sons of Mary: Forging iron and finding gold in engineering and business ideologies. In S. H. Christensen, B. Delahousse, C. Didier, M. Meganck, & M. Murphy (Eds.), *The engineering-business nexus: Symbiosis, tension and co-evolution* (Vol. 32, pp. 153–172). Springer. https://doi.org/10.1007/978-3-319-99636-3_8

Lea, N. (2019). *Lady with the iron ring: An engineer's memoir of hope, luck and success*. Friesen Press.

Levert, D. (2020). *On cold iron: A story of hubris and the 1907 Quebec bridge collapse*. Friesen Press.

Leydens, J. A., & Lucena, J. C. (2018). *Engineering justice: Transforming engineering education and practice*. John Wiley & Sons; IEEE Press.

Lönngren, J., Adawi, T., & Berge, M. (2020). "I don't want to be influenced by emotions"—Engineering students' emotional positioning in discussions about wicked sustainability problems. *2020 IEEE Frontiers in Education Conference (FIE)*, 1–5. https://doi.org/10.1109/FIE44824.2020.9273946

Macdonald-Roach, E., Rottmann, C., Moore, E., Chan, A., Radebe, D., Van Beers, S., & Nixon, S.-A. (2024). Are hardhats required for engineering identity construction? Gendered and racialized patterns in Canadian engineering graduates' professional identities. *2024 ASEE Annual Conference & Exposition Proceedings*, 46596. https://doi.org/10.18260/1-2%2D%2D46596

Martin, T., & Hoffman, S. M. (2011). *Power struggles: Hydro development and First Nations in Manitoba and Quebec*. University of Manitoba Press.

Maslen, S., & Hayes, J. (2020). "This is how we debate": Engineers' use of stories to reason through disaster causation. *Qualitative Sociology, 43*(2), 191–212. https://doi.org/10.1007/s11133-020-09452-1

McLean, M., Khaira, A., & Alexander, C. (2023). Symbols and rituals are alive and well in clinical practice in Australia: Perspectives from a longitudinal qualitative professional identity study. *Medical Teacher, 45*(12), 1425–1430. https://doi.org/10.1080/0142159X.2023.2225722

Merton, R. (1957). *Social theory and social structure*. Free Press.

Milard, J. R. (1988). *The master spirit of the age: Canadian engineers and the politics of professionalism*. University of Toronto Press.

Ng, E., Fitzsimmons, T., Kulkarni, M., Ozturk, M. B., April, K., Banerjee, R., & Muhr, S. L. (2025). The anti-DEI agenda: Navigating the impact of Trump's second term on diversity, equity and inclusion. *Equality, Diversity and Inclusion: An International Journal, 44*(2), 137–150. https://doi.org/10.1108/EDI-02-2025-0116

Parry, B. (1996). The content and discontents of Kipling's imperialism. In *Reading Fin de Siècle Fictions*. Routledge.

Paul, R. M., Zacharias, K., Nolan, E. M., Monkman, K., & Thomsen, V. (2023). Stubborn boundaries: The iron ring ritual as a case of mapping, resisting, and transforming Canadian engineering ethics. *Frontiers in Education, 8*, 1177035. https://doi.org/10.3389/feduc.2023.1177035

Perry, A. (2016). *Aqueduct: Colonialism, resources, and the histories we remember*. ARP Books.

Petroski, H. (1995). Engineering: The iron ring. *American Scientist, 83*(3), 229–232.

Prasad, A., & Śliwa, M. (2024). Critiquing the backlash against wokeness: In defense of Dei scholarship and practice. *Academy of Management Perspectives, 38*(2), 245–259. https://doi.org/10.5465/amp.2023.0066

Radebe, D., Rottmann, C., Chan, A., Macdonald-Roach, E., & Moore, E. (2023). Who's experiencing weathering? A question of belonging in the engineering profession. *Proceedings of the Canadian Engineering Education Association (CEEA)*. https://doi.org/10.24908/pceea.2023.17158

Riley, D. (2008). Ethics in context, ethics in action: Getting beyond the individual professional in engineering ethics education. *Peer.* 13.570.1–13.570.18. https://peer.asee.org/ethics-in-context-ethics-in-action-getting-beyond-the--individual-professional-in-engineering-ethics-education

Riley, D. M., & Lambrinidou, Y. (2015). Canons against cannons? Social justice and the engineering ethics imaginary. Peer. 26.322.1–26.322.19. https://peer.asee.org/canons-against-cannons-social-justice-and-the-engineering-ethics-imaginary

Riley, D. M., Slaton, A. E., & Herkert, J. R. (2015a). What is gained by articulating non-canonical engineering ethics canons? *Peer.* 26.1723.1–26.1723.16. https://peer.asee.org/what-is-gained-by-articulating-non-canonical--engineering-ethics-canons

Riley, D., Slaton, A., & Pawley, A. (2015b). Social justice and inclusion: Women and minorities in engineering. In *Cambridge handbook of engineering education research* (pp. 335–356). https://doi.org/10.1017/CBO9781139013451.022

Roeser, S. (2012). Emotional engineers: Toward morally responsible design. *Science and Engineering Ethics, 18*(1), 103–115. https://doi.org/10.1007/s11948-010-9236-0

Rottmann, C., Reeve, D., Sacks, R., & Klassen, M. (2018). Where's my code? Engineers navigating ethical issues on an uneven terrain. *2018 ASEE Annual Conference & Exposition Proceedings*, 31242. https://doi.org/10.18260/1-2%2D%2D31242

Rottmann, C., & Reeve, D. (2020). Equity as rebar: Bridging the micro/macro divide in engineering ethics education. *Canadian Journal of Science, Mathematics and Technology Education, 20*(1), 146–165. https://doi.org/10.1007/s42330-019-00073-7

Rottmann, C., Moore, E., Chan, A., & Radebe, D. (2021). Who can I turn to? Engineers navigating ethical dilemmas at work. *2021 IEEE International Symposium on Ethics in Engineering, Science and Technology (ETHICS)*, 1–5. https://doi.org/10.1109/ETHICS53270.2021.9632801

Secules, S. (2019). Making the familiar strange: An ethnographic scholarship of integration contextualizing engineering educational culture as masculine and competitive. *Engineering Studies, 11*(3), 196–216. https://doi.org/10.1080/19378629.2019.1663200

Seniuk Cicek, J., Paul, R. M., Martin, D. A., & Riley, D. (2025). A feminist critical standpoint analysis of engineering ethics education and the powers at play in accreditation, research, and practice. In *The Routledge International Handbook of Engineering Ethics Education*. Routledge.

Seron, C., Silbey, S., Cech, E., & Rubineau, B. (2018). "I am not a feminist, but...": Hegemony of a meritocratic ideology and the limits of critique among women in engineering. *Work and Occupations, 45*(2), 131–167. https://doi.org/10.1177/0730888418759774

Slaton, A. E. (2012). Engineering improvement: Social and historical perspectives on the NAE's "Grand Challenges.". *International Journal of Engineering, Social Justice, and Peace, 1*(2). https://doi.org/10.24908/ijesjp.v1i2.4305

Slaton, A. E., Cech, E. A., & Riley, D. M. (2019). Yearning, learning, and earning: The gritty ontologies of American engineering education. In *STEM of desire* (pp. 319–340). https://doi.org/10.1163/9789004331068_017

Smith, J. M., Beddoes, K., Downey, G., Jesiek, B. K., Reddy, E., Riley, D., Rossmann, J. S., Wylie, C. D., York, E., da Costa Marques, I., Konstantis, K., & Mody, C. C. M. (2025). We have been here before: Reflections on engineering and authoritarianism. *Engineering Studies, 17*(1), 1–10. https://doi.org/10.1080/19378629.2025.2466994

Song, Y. K. (2018). Rituals in medicine: The morbidity and mortality conference. *Journal of Ritual Studies, 32*(1), 1–10.

The Corporation of the Seven Wardens. (2022, November 21). *Seven Wardens announce review of the ritual of the Calling of an Engineer*. https://ironring.ca/wp-content/uploads/2022/11/Communique-_EN-20221121.pdf

The Corporation of the Seven Wardens. (2025, January 13). *Seven Wardens approve modernized "Calling of an Engineer"*. https://ironring.ca/wp--content/uploads/2025/01/Communique-Obligation-Ceremony-Jan-2025_EN_final.pdf

The Corporation of the Seven Wardens. (n.d.). *Frequently asked questions: The ritual of the Calling of an Engineer*. Retrieved November 13, 2024, from https://ironring.ca/faq-en/

The Iron Ring Podcast. (n.d.). *Women in engineering—Concordia University*. Retrieved November 13, 2024, from https://www.womeninengineeringconcordia.com/the-iron-ring-podcast

The Ritual of the Calling of an Engineer. (n.d.). *Camp 18*. Retrieved November 13, 2024, from https://camp18ironring.com/

Tonso, K. L. (2014). Engineering identity. In A. Johri & B. M. Olds (Eds.), *Cambridge handbook of engineering education research* (1st ed., pp. 267–282). Cambridge University Press. https://doi.org/10.1017/CBO9781139013451.019

Tuck, E., & Gaztambide-Fernández, R. A. (2013). Curriculum, replacement, and settler futurity. *Journal of Curriculum Theorizing, 29*(1), 72–89.

Zacharias, K., Hersh, M., Brightman, A., & Beever, J. (2025). Individual and collective dimensions of ethical decision-making in engineering. In *The Routledge international handbook of engineering ethics education*. Routledge.

CHAPTER 9

Industry and Geopolitical Tensions: Ethical Dilemmas When Safety Meets Security

Sissel Haugdal Jore, Ole Andreas Hegland Engen, Susanne Therese Hansen, and Øyvind Midttun

1 Introduction

> [O]ur ideas about how threats should be defeated and who is responsible for this fight, have fundamentally changed. Today's geopolitics are fought not only by states, within the framework of diplomacy and the military institutions, but also by private companies. (Karen Lund Petersen (2023)

The new geopolitical era after Russia's war on Ukraine has fundamentally changed the role of the Norwegian petroleum industry and its workers. In

S. H. Jore (✉) • O. A. H. Engen
University of Stavanger, Stavanger, Norway
e-mail: sissel.h.jore@uis.no; ole.a.engen@uis.no

S. T. Hansen
NTNU Social Research, Trondheim, Norway
e-mail: susanne.hansen@samforsk.no

Ø. Midttun
Norwegian Ocean Industry Authority, Stavanger, Norway
e-mail: oyvind.midttun@havtil.no

J. Hayes, S. Maslen (eds.), *Ethics in the doing of technoscience*,
https://doi.org/10.1007/978-3-032-20002-0_9

the months following Russia's full-scale invasion of Ukraine in February 2022, it became increasingly clear how the Norwegian petroleum industry was in an entirely new threat situation. A series of sightings of what were referred to as "unidentified drones" at offshore and onshore petroleum facilities attracted public attention and created uncertainty and fear among those working in the sector (Norwegian Ocean Industry Authority, 2022). The situation was further exacerbated by the Nord Stream sabotage in the Baltic Sea in September 2022 which resulted in several underwater explosions and gas leaks on the Nord Stream 1 and 2 pipelines built to transport Russian natural gas to Germany via the Baltic Sea. The explosions were located in international waters, between the economic zones of Denmark and Sweden. As described by Moe and Hansen (2025, pp. 389–90), the Nord Stream pipeline sabotage is confirmed to be a deliberate act but remains officially unsolved, although investigations are ongoing. In the aftermath of the sabotage, European leaders and NATO have routinely pointed at Russia as the most likely perpetrator. The sabotage of the Nord Stream pipelines garnered significant attention from the media, national governmental authorities, European nations, and NATO regarding the potential for energy supplies to be targeted by Russia as a tactic of hybrid warfare. In Norway, the pipeline sabotage resulted in Norwegian petroleum sector becoming significantly concerned about the safety and security of their employees. Continuous reports of drones near Norwegian offshore installations further amplified this anxiety, highlighting the industry as a potential target in the geopolitical conflict. The Norwegian Ocean Industry Authority—the supervisory authority for industrial activity on the Norwegian continental shelf—explicitly urged petroleum companies to be more vigilant, warning that drones near offshore installations could pose a risk of accidents or deliberate attack (Norwegian Ocean Industry Authority, 2022).

In light of apprehensions regarding possible threats to its petroleum infrastructure, Norway heightened military and naval surveillance and presence around offshore oil and gas facilities and stationed armed troops to safeguard essential sites on land. The government also subjected key petroleum companies to the national Security Act, giving companies explicit responsibilities for taking national security into consideration in their risk assessments. These measures underscored the perception of geopolitical tensions causing an increased threat to workers in the sector.

In this chapter we explore the ethical challenges and compromises faced by the petroleum sector and its workforce following the sabotage of the Nord Stream gas pipelines. The chapter uses document analysis of the minutes of the Safety Forum meetings and interviews with members of the collaborative tripartite platform of Safety Forum. We examine how changes in the threat environment and increased focus on security have influenced the collaborative practices between authorities, employers, and employees within the institutionalized tripartite model of the petroleum industry. Doing so, we scrutinize the communication and dialogues within and between the political, administrative, and legal structures of the Norwegian tripartite system, to detect how authorities, industry, and the labor unions strived to balance different interests while adjusting to new external requirements.

This chapter illustrates how geopolitical shifts and the associated role of companies as security actors challenge the core values and identity of working life in a liberal democracy. We conclude that the transformed geopolitical landscape comes with new risks that skew the balance between safety and security work within the sector and introduce new ethical dilemmas for the organizations involved. An ethical dilemma arises when one is faced with conflicting moral duties, where fulfilling one may come at the expense of another. In the context of safety and security, such dilemmas involve balancing the need to protect people or systems with the responsibility to uphold rights, transparency, and public trust. These challenges may be perceived differently from the viewpoints of authorities, corporations, and employees. The tripartite cooperation model has demonstrated resilience and adaptability, but security concerns have reshaped power dynamics, with employers increasingly aligning with state security priorities.

2 Norwegian Petroleum Companies and Their Workers as Security Actors in the New Geopolitical Era

As the above quote illustrates, current geopolitics are fought not only by states. Companies nowadays play a pivotal role as security policy actors (Petersen, 2023). The current era is defined by rapid technological advancement, shifting power dynamics, and heightened global tensions. The convergence of artificial intelligence, terrorism, hybrid threats, and geopolitical tensions has created unprecedented challenges for both states and companies (Storesund et al., 2024). This new geopolitical landscape has created a new category of risk problems for private companies that creates new challenges for security and risk management in private companies and dissolve the traditional relationship between the state and private enterprises (Hansen & Antonsen, 2024; Skare & Jore, 2024). According to Petersen (2023), the role of private companies has changed along four different dimensions following altered geopolitical framework conditions: (1) the company as a critical infrastructure owner, (2) the company as a tool in the security policy of the state, (3) the company as an independent security actor, and (4) the company as a symbol of Western values. In the following, we detail how each of the four apply to the case of the Norwegian petroleum industry.

Firstly, the private sector owns about 80% of critical infrastructure such as energy, transport, water, food, which is vital for society to function in peace, crisis, and wartime (Petersen, 2023). This has led to a form of privatization of security where private companies co-produce security alongside states, leveraging their economic power and technological expertise to influence security governance. This makes private businesses an extension of states' national security interests, because they provide services or goods that modern societies and accordingly states rely on to maintain peace and stability. The implication is that organizations and their workers become extensions of the state's national security apparatus, especially in times of heightened threat or crisis. The Norwegian petroleum sector remains a case in point. The sector is deemed vital to national security interests due to its economic, geopolitical, and strategic significance (Norwegian Ocean Industry Authority, 2023). The sector is Norway's largest revenue source and a major source of funding for the welfare state. In addition, as Europe's largest natural gas supplier after Russia's full-scale invasion of Ukraine, Norway ensures stable energy supplies to the European Union and the UK, reducing continental European countries' reliance on Russian energy resources. Norway's petroleum infrastructure is now interconnected to European energy security and global market stability, increasing the sector's geopolitical and strategic importance (Hansen & Moe, 2025).

In Norway, the increased strategic role of the sector after the invasion of Ukraine and events like Nord Stream has been followed by subjecting key actors in the sector to domestic security regulation. In 2022, the petroleum companies Equinor and Gassco were placed under the jurisdiction of the national Security Act. With this, they became responsible for analyzing, assessing, and dealing with intentional actions that may "harm national security interests" (Security Act, 2019). By subjecting companies to this law, the sector and its workers became not only a provider of critical infrastructure and energy, but also a co-provider of national and European security. The practical implication for workers is that they have been subjected to stricter security measures, including background checks, access controls, and monitoring, to prevent insider threats or sabotage.

Second, private companies have now become an offensive tool in the security policy of the state. This is typically illustrated with the Chinese company Huawei's investment in critical technology in Western countries or SOE Gazprom acting as the "arm of the Kremlin" in pursuit of foreign policy goals (Boute, 2022). We will not resort to arguing that the Norwegian government is using the petroleum industry as a security policy tool. However, since the Norwegian petroleum sector has become increasingly critical to both Norwegian and European security, there is always the chance that an actor like Russia may see the sector as an offensive tool of the Norwegian state, the European Union, and NATO. Norway is now Europe's largest gas supplier, replacing Russia's long dominant stronghold on the European energy market. Meanwhile, Norwegian energy infrastructure has consequently become a target for cyberattacks, espionage, and sabotage. The Norwegian Intelligence Service's annual threat assessment accordingly places great emphasis on the potential consequence (Norwegian Intelligence Service, 2024):

> Currently, Norway is a key supplier of gas to Europe. Norwegian gas fields and terminals are directly connected to receiving countries in Europe through a vast pipeline network. A significant share of the gas consumed in Germany, UK, Belgium and France comes from Norway. The gas infrastructure could be subject to accidents, physical sabotage and destructive cyberattacks. Damage to Norwegian petroleum infrastructure would harm both Norway and the receiving countries in Europe.

Russia has been mapping Norwegian critical petroleum infrastructure for years. This mapping is still ongoing, both physically and in the digital domain. The knowledge acquired through the mapping could become strategically and operationally important in a conflict situation. According to the 2024 National Threat Assessment report of the Norwegian Police Security Service, Russia views "energy-related measures as a key part of making friction in the West" (Norwegian Police Security Service, 2024). Possibly, Russian sabotage "against targets related to Norwegian gas exports may be intended to trigger or reinforce an energy crisis in Europe" to weaken support for Ukraine (Norwegian Police Security Service, 2024, p. 10). The focus is on Russia which remains the most significant threat due to its war in Ukraine, Russo-NATO tensions, and energy-related interests. However, the Norwegian Intelligence Service also identifies

China, Iran, and North Korea as key intelligence threats. China poses a growing threat linked to Arctic ambitions and control over supply chains. Iran and North Korea are active in targeting Norwegian interests through cyber operations and espionage. Regardless of the adversary, the sector faces threats like sabotage and cyberattacks from state actors (or non-state actors operating on their behalf). This makes workers and their activities a gray zone between commercial and governmental interests directly tied to national security outcomes, making workers potential targets of state-sponsored espionage, sabotage, or cyberattack.

Third, private companies can act as independent security actors. For instance, private tech giants like Microsoft, Amazon, and SpaceX have provided Ukraine with critical digital infrastructure, cloud services, and communication tools, aiding in cyber defense and intelligence gathering at the behest of governments. Many companies do not act entirely by themselves but play a role in a broader political game, where regardless of company motive, they have become powerful political negotiators of what can be considered legitimate security policy.

Again, Norwegian petroleum companies remain a case in point. The industry has been impacted by sanctions on Russia, including bans on the import of Russian oil, as well as price caps on petroleum products. These measures have reduced global oil and gas supply from Russia, driving up oil and gas prices and creating market disruptions. Norwegian companies have also ceased trade with Russian counterparts, further limiting economic interactions. Additionally, sanctions have curtailed cooperation in Arctic oil and gas exploration, affecting long-term strategies for both nations in the Barents Sea. Although these examples are related primarily to sanctions and state policies, Norwegian companies are used to taking a social responsibility. Yet there have also been instances where petroleum companies have faced criticism for choosing profit over security. A relevant example is the Norwegian petroleum company Equinor, which entered into a substantial agreement with the Chinese drilling company COSL for the use of floating rigs, valued at approximately $369 million. The contract includes energy efficiency upgrades that align with Equinor's strategy for safe and sustainable operations on the Norwegian continental shelf (Equinor, 2025). However, this deal has sparked concerns in Norway over potential security risks due to COSL's ties to Chinese state interests. In sum, the above illustrates how private companies operating in strategically important markets are not protected from, or unaffected by, global geopolitics. Accordingly, employees may find themselves contributing to decisions that have broad political and societal impacts.

Fourth, many companies today have become national symbols, directly linked to national security policy orientation. This was evident in connection with the war in Ukraine, where Norwegian companies' presence in Russia led to accusations of non-patriotism and of compromising Norwegian security and allied commitments. The activities of petroleum companies were thus no longer politically neutral. Rather, companies like Equinor were forced to take a security policy position to the war and to its own involvement in Russia, to bring activities into alignment with Norwegian security policy and the European morale.

Companies, their infrastructures, and activities are important political symbols not only for their flag state, but also for the actor seeking to weaponize company assets. Indeed, it is highly likely that a potential sabotage against a Norwegian gas pipeline will be intended and interpreted as a form of hybrid warfare against Norway. An attack on Norwegian petroleum infrastructure could potentially be interpreted as an attack against NATO too, since Norway is a NATO member, and its energy infrastructure is critical to European energy security (RAND, 2024). This could potentially trigger mutual defense obligations under Article 5 of the NATO Pact—an outcome the potential of which is no secret to hostile actors. The consequences are that Norwegian petroleum companies nowadays must navigate a complex landscape of public scrutiny, ethical challenges, operational uncertainty, and heightened security risks.

Overall, the Norwegian petroleum industry now plays multiple roles as security policy actors, and these have direct consequences for many of its workers. However, these roles do not come without dilemmas. The security policy pursued by large companies is not based on joint democratic political decisions or a Norwegian political strategy. Moreover, the sector's newly acquired security policy roles might be perceived in various ways by different stakeholders involved.

3 Theoretical Framework

The Norwegian petroleum industry assumes four distinct security policy roles—described above—each presenting ethical dilemmas, compromises, and trade-offs for the sector as it transitions into a new position as an actor within international geopolitics and national security. The evolving role of petroleum companies signifies a shift that places a responsibility for national security upon organizations; this is a responsibility that they may not be inherently equipped or organized to address (Skare & Jore, 2024).

The evolving roles of petroleum companies will present significant challenges for both the organizations and their workforce. This is primarily because geopolitical tensions have established a new framework condition necessitating a more prominent international aspect and a stronger security dimension in the governance of risks associated with the petroleum industry. Particularly important for our purpose, there will be shifts in the dynamics and equilibrium between safety and security, both within the petroleum companies themselves and in their interactions with governmental bodies and influential international entities, including nation-states and the industry at large (see also Hansen & Antonsen, 2024).

3.1 Safety and Security

The concepts of safety and security are often used to distinguish between separate fields of risk handling. Cyber risks, terrorism, and other intentional malicious crimes are associated with security, while safety implies protection from unintentional acts where human malicious intent is not present (Jore, 2023). Safety and security refer to phenomena with differences in ontology (intended and malicious vs. no malicious intent), epistemology (the ways of creating knowledge about risk problems), practice (the professional knowledge involved in applying the acquired knowledge), and communication (openness in safety vs. secrecy in security) (Hansen & Antonsen, 2024).

From an organizational perspective, these risks are fundamentally different in nature (Jore, 2019). Safety risks are associated with an organization's production and profit. Production of goods and services is always connected with some kind of risk, and these are risks an organization is willing to take to produce its desired outcome and to gain profit. The sources of these risks are generally well-known, and an organization can often use reliable historical data in the risk management process. Since organizations have knowledge concerning the risks, they usually also know how these risks can be mitigated. Safety rests on the presumption that the involved parties have a common interest in avoiding an incident. To this end, for safety information sharing, transparency, dialogue, and trust are considered crucial.

Conversely, security involves the threats to which organizations are exposed. Security risks are not necessarily directly linked to the production of an organization and are, therefore, less controllable from an organizational perspective (Jore, 2019). Since the nature of the threat often is rooted outside the organization, most organizations do not have the means to fully understand and reduce the threat. Creating knowledge about security risks often involves intelligence and draws upon intelligence services. Moreover, organizations will, in many cases, also lack the means to reduce security threats since it is the state that has the mandate to discover and arrest criminals, spies, terrorists, hackers, or other possible perpetrators. Security risks are characterized by a malicious and often criminal intent. However, in contemporary geopolitics, legal and democratic measures may also be applied by hostile states. Examples include foreign interference and foreign acquisition.

Safety and security have different ethical implications. Safety is most often associated with trust, transparency, openness, dialogue, and information sharing, often perceived as low politics and bureaucratic procedures. Security threats are associated with matters of high politics related to crime prevention, national security, and state sovereignty and have traditionally been dealt with within the domains of the police and defense policy. Security is related to secrecy and information sharing within trusted communities. Consequently, companies that predominantly deal with safety risks can be expected to be significantly challenged when having to deal with security risks.

3.2 *Risk and Ethics*

All decisions involving risks, whether risks can be categorized as safety risks or security risks, are influenced by values and ethical considerations. Ericson and Doyle, for example, explored the intersection of risk and moral principles and illustrate how risk discourse shapes social behavior and moral regulation (Ericson & Doyle, 2003). The literature on risk and ethics often describes risk management not only as a practical concern but also as a phenomenon that enforces societal norms and values. Thus, the intersection of risk, ethics, and morality in modern society involves ethical evaluations of actions that impose risks on others. Additionally, risk is often moralized through societal norms, requiring individuals or institutions to self-regulate their behavior to align with shared values of safety, security, and responsibility. Institutions such as the tripartite model in the Norwegian petroleum industry also face ethical challenges in managing risks, such as balancing rules-based compliance with values-based leadership, especially in uncertain environments like climate change or geopolitical instability. Additionally, risk discourses can normalize certain behaviors and shape perceptions of what constitutes moral or immoral actions within specific contexts as they guide the evaluation of potential harm, fairness, and responsibility. Since risk is related to values, risk perception research has highlighted how significant differences exist in how individuals, groups, and organizations perceive risks. Ethics is often framed as responsibility or a matter of how to assign blame or liability in cases of failure or an accident (Maslen et al., 2021). However, ethics as responsibility has also been addressed in the literature as forward-looking and preventative (Harris, 2008; Van de Poel, 2011).

3.3 Norway's Tripartite Model: Balancing Safety, Security, and Trust

A fundamental building block of the regulatory regime in the Norwegian petroleum industry is that "each individual company is responsible for the safety and industrial security of their own activities" (PSA, 2017). This logic of decentralizing responsibility through enforced self-regulation is now also applicable for security (PSA, 2017, 2022). This regime rests on the assumption that the involved parties have a common interest in maintaining the system and that the conflicts of interest that may arise when doing so will naturally be solved within and between involved actors without threatening the foundation of trust between the involved parties (Engen et al., 2023).

In the petroleum industry, deliberation and cooperation through the regular collaborative tripartite meeting arena of "Safety Forum" plays a central role in the industry's efforts to reduce occupational and major accident risks. Consequently, in this book chapter we scrutinize the communication between the political, administrative, and legal structures that constitute the Norwegian tripartite system in the Safety Forum, to detect how authorities, industry, and labor unions strived to balance different interests while adjusting to a new threat situation after the Nord Stream pipeline sabotage.

The tripartite model reflects shared Nordic principles and provides a robust framework for navigating an increasingly complex and dynamic landscape of security and safety. By prioritizing equality, clearly defining responsibilities, and emphasizing local-level management, the model ensures that conflicts related to risk regulation are addressed equitably and effectively. Moreover, its focus on collaboration fosters strong cooperation among stakeholders, enhancing the industry's ability to respond to and recover from a wide range of security challenges.

This tripartite model embodies a broader commitment to governance grounded in shared values and democratic ideals. It underscores the importance of societal resilience, highlighting the critical role of community and governance in upholding democratic integrity—not only in the workplace but also across society. The model illustrates the interplay between material processes and deeper ethical, cultural, and societal dimensions of safety and security. It proposes that safety and security are not merely about safeguarding infrastructure or implementing technical measures; rather, it encompasses a broader understanding of how societal values, community cohesion, and collective identity shape both the perception of threats and the strategies used to address them within workplaces, industries, and society in general (Engen et al., 2017; Forseth & Rosness, 2021).

On a practical level, regulatory discussions within the tripartite framework address physical infrastructure, logistical measures, and bureaucratic systems, framing security as a technical issue managed through emergency responses, critical infrastructure protection, and coordinated actions. However, the ethical dimensions of these discussions emphasize values such as democracy, openness, and compassion, prioritizing resilience and the protection of workers.

The dual role of the tripartite model—managing both technical and ethical dimensions—highlights its vulnerability to emerging threats while also serving as a source of cohesion, solidarity, and security. It reflects the interdependence between structural mechanisms and intangible societal qualities, as many threats arise from tensions inherent in liberal societies' value systems.

The traditional openness of the tripartite cooperation has been challenged by the increasing need for confidentiality and secrecy due to new security concerns. This shift places national security responsibilities on the tripartite model itself, shaping the industry's approach to risk governance and requiring careful balancing of transparency and protection.

The changes in the threat landscape after the Russian invasion of Ukraine, the drone activities, and the Nord Stream sabotage altered the responsibilities of the Norwegian petroleum industry and triggered concerns about the safety and security of the workers. This development may significantly impact established labor relations among the authorities, companies and among the employees in the industry. The Norwegian petroleum sector is the workplace for thousands of people whose safety may be compromised by security threats currently and in the future. The new role of industry, as a security provider with specific responsibilities under national security legislation, might threaten the trust and involvement between different parties in the industry. This could potentially have major implications also for the safety regime that is built on trust and collaboration between the different parties involved.

4 Methods and Data

This chapter is a qualitative case study presenting a case of ethical consequences for the workforce following changed geopolitical framework conditions and specific attacks. Deliberations and interviews in the Safety Forum provide an excellent example of how hybrid threats targeted at a particular industry generate new ethical dilemmas. While our findings represent only one case and are specific to the Norwegian setting and to the maritime energy industry, they have potency in enticing relevant anticipations about possible ethical dilemmas in similar cases in other geographies and sectors facing hybrid threats.

The empirical material and core analytical insights presented in this chapter are based on a qualitative study including 14 interviews with members of the Safety Forum and document analysis of 10 Safety Forum meeting minutes from March 2022 to March 2024 (Midttun, 2024). The material has been re-analyzed and expanded in light of broader ethical and geopolitical considerations. We have focused on how the new security situation and changes in the threat landscape affected established practices related to openness, information sharing, and trust. The main emphasis of the empirical material lies in the interviews, although the meeting minutes from the Safety Forum also provided relevant information for this study.

The informants for the interviews were selected among the members of the Safety Forum. Most members of the Safety Forum have extensive industry and organizational experience, and the majority also have longstanding involvement in the Safety Forum. They are therefore well-acquainted with the industry and the role of the Safety Forum. Representation and relevance were important selection criteria. The Safety Forum consists of a total of 23 members. Through the selection of informants, representation from all member organizations (employer organizations and trade unions), as well as the Norwegian Ocean Industry Authority (Havindustritilsynet [Havtil]), was ensured. The interviews focused on how the members of the Safety Forum experienced the increased attention to security in the Safety Forum, what challenges the representatives in the Safety Forum perceived that the increased focus on security posed for collaborative practices, to what extent the representatives accepted the increased focus on security in the Safety Forum, and, if so, what were the main factors underlying this acceptance.

The minutes from the Safety Forum meetings were downloaded from the web page of the Norwegian Ocean Industry Authority. These minutes were analyzed to examine how changes in the threat landscape were discussed and how security-related topics were incorporated into the Safety Forum's agenda during the relevant period. This analysis helped to supplement and support the information obtained from the interviews and provided important insights into the discourse on security within the Safety Forum—specifically how, when, and in what manner security topics were addressed.

5 Findings

5.1 Consensus on the Industry as a Security Policy Actor and Safety Forum as a Security Arena

Following the Russian invasion of Ukraine in 2022, the agenda of the Safety Forum has increasingly incorporated topics pertaining to security. This has marked a shift and a potential challenge for cooperation between the parties involved, as security by nature is more associated with confidentiality and protection than with openness and trust (Jore, 2019, 2023; Piètre-Cambacédès & Chaudet, 2010).

The parties involved in the tripartite cooperation did not question that the geopolitical unrest had become a major framework condition after the Nord Stream pipeline sabotage. It was a conscious choice to use the Safety Forum to discuss issues regarding security, because the Safety Forum was seen as an arena for information sharing and the perceived seriousness of the situation made fast information sharing important. One of the government representatives in the Safety Forum said that there was no doubt that the Safety Forum was the right arena to use, even though uncertainty initially reigned about how security would be handled in the tripartite cooperation: "In the beginning, there was uncertainty about what this situation was and how we should handle it. At the same time, we were very clear that the Safety Forum was the right arena instead of creating an alternative arena."

The representative described how an extraordinary meeting was held in the Safety Forum to ensure that information was provided to its members and to get feedback on what the parties were concerned about and how they acted.

This view was shared by the other parties in the tripartite cooperation. A union representative claimed that the use of the Safety Forum as an arena for sharing information about security meant that the authorities quickly shared information to the parties after the Nord Stream incidents and drone sightings:

> Yes, in that situation I think [Safety Forum was] very good. I have to say that. Because [Safety Forum was] very quick to call for information meetings and bring in expertise on security [...] and things like that. Quite quickly, so that we felt that we had an arena where we could air our concerns [about drones].

The representative also described how security was raised as a topic at the extraordinary Safety Forum meeting where also external experts were invited to share information, and he believed that the authorities had understood the importance of rapid information sharing. Following several security-related incidents affecting workers, union representatives have used the opportunity to present their views in the Safety Forum.

5.2 Security as a Challenge to Openness and Information Sharing

Despite the consensus on utilizing the Safety Forum as an arena for information sharing, there have been notable changes in the tripartite institutions, particularly in the areas of openness and information sharing resulting in growing tensions among authorities, employers, and employees in the petroleum sector. These tensions primarily stem from the inherent variations in openness and information-sharing practices between safety- and security-focused goals.

Vignette 1 describes a typical Safety Forum conversation about a safety issue. The process as described illustrates the effectiveness of the tripartite safety model, where all actors—regulators, employers, and unions—have a voice in shaping what constitutes acceptable risk.

In contrast, Vignette 2 describes a Safety Forum conversation regarding a specific security issue. It raises questions about transparency, inclusion, and how the concept of security risk might be broadened to acknowledge frontline experiences.

Box 9.1 Vignette 1: Safety—A Model of Tripartite Collaboration

Normally during maintenance shutdown on an offshore installation, operators propose to reduce the size of the on-site crew and automate more inspection routines using drones and remote sensors. The proposal is brought to the Safety Forum, where union representatives immediately raise concerns. They argue that physical presence is vital for catching weak signals, maintaining situational awareness, and responding quickly to unforeseen events. Drawing on the established tripartite model, the unions can influence the risk assessment process, emphasizing a "zero harm" approach over ALARP-based justifications. The result is a negotiated compromise: certain drone inspections are approved, but manned inspections are retained in critical areas. Additional training is provided for new technologies, and workers remain actively involved in monitoring routines.

Box 9.2 Vignette 2: Security—A Less Settled Space for Worker Involvement

In response to rising geopolitical tensions and drone sightings near offshore infrastructure, a company may introduce new digital surveillance tools and update its cybersecurity protocols. These changes are implemented rapidly, driven by national-level security advisories and expert risk assessments. While management views these actions as necessary and time-sensitive, offshore workers express unease: they notice new monitoring systems but have received limited information or training, and there is uncertainty about how these changes affect their roles and responsibilities.

At the Safety Forum, security issues are discussed, but the dialogue is more technical and centralized than for traditional safety topics. Union representatives support strong security measures in principle—as reflected in worker comments about wanting to "defend the platform"—but feel sidelined in defining what "security" means in practice. Unlike with safety, where their participation is well institutionalized, the role of workers in shaping security policy remains underdeveloped.

A key difference in these conversations is the level of information that can be shared. Against a backdrop in which, for safety, openness is seen as best-practice, a government representative said it was initially difficult to find out how much information about security could be shared. This representative was of the view that in the beginning they were being more restrictive than necessary about sharing information. This led to some representatives being left with the impression that the information was unevenly distributed or withheld.

> We struggled a bit to find the level of what kind of information we could provide. It was quite often that we said that we couldn't share, we couldn't share everything. It created a somewhat demanding situation, with people feeling that there was a lot they weren't told. And that actually became difficult, because we didn't provide that much information, but we gave the impression that we did. And of course we provided some more information, but the impression was that the employee side in particular didn't have the information they thought they should have.

In light of this uncertainty, Safety Forum representatives sought guidance on the appropriate level of information sharing. When asked, security representatives often advised: "Share as much as you can and keep secret as much as necessary." However, this guidance is vague and unsatisfactory from the standpoint of sound risk and safety practices. It is ethically problematic from a safety perspective if important information about workforce conditions is withheld—especially if it later turns out that the information was not actually subject to security classification.

The Safety Forum has an established culture of sharing, aligned with tripartite traditions, and is built on trust. However, the Safety Forum's increased focus on security has activated latent tensions, challenging these established patterns of interaction. Over time, openness and sharing of information about security in the Safety Forum moved from an initial phase characterized by uncertainty and restraint to a situation of normalization and gradually more openness and sharing.

5.3 *Changes to Flows of Information in Response to Security*

Despite the tensions caused by the increased focus on security, the members of the Safety Forum demonstrated both understanding and acceptance of limited openness, recognizing that the rules of engagement had changed because of geopolitical tensions. While there were collective expectations for transparency, the members accepted that interaction on security could not be performed with the same degree of openness as for safety issues. One employer representative believed that from the employee side there had been a lack of access to information about the security situation, and that this contrasted with the openness that previously had been the norm in the Safety Forum:

> They are used when you work with HSE in all facets, in a way trust and openness are absolutely fundamental, there is complete openness about things. And then the employee representatives probably notice that when it comes to such security matters, when it really comes down to it, there are in a way some other rules of the game and mechanisms that come into play.

However, this was not simply a matter of the degree of information sharing. This representative pointed out that information exchange about security took place through several other channels and forums than the Safety Forum. These channels did not necessarily have the same employee representation as in the Safety Forum. This could result in an unbalanced distribution of information and could lead to frustration and a lack of a joint understanding and situational awareness. Security clearance was also raised as a problematic area:

> Some people have security clearance, and some have not, and then [security] matters are of a different nature. And in that case, it is often easier for the industry to have closer and more open contact with the authorities, including the Norwegian Ocean Industry Authority, than it is to have the same degree of openness in relations with employee representatives. So they do that.

One employer representative described how it was challenging to practice transparency about security in a tripartite context, and that it was often easier for the employer to have an open and unfiltered exchange of information with the authorities than with employee representatives. Although this created frustration on the employee side, the informant did not see how the challenge could be easily solved:

> It's not about a lack of trust in the employee representatives, but it's a bit of the nature of the whole [matter] that it's easier to talk openly and directly and without a filter with the authorities about this. (...) I've probably experienced that the employee representatives have expressed a certain kind of frustration with regard to the lack of transparency. But I don't quite see

> how it could have been done differently. Because there are things in this that you can't spread information about so openly.

Some of the employee representatives thought the solution to this was security clearance. However, according to an employer representative, security clearance did not necessarily make information sharing easier, since classified information cannot be shared further:

> Then it will just be about providing information that they cannot share anyway. It is difficult to find any practical reason why it would make any difference if you gave this information to a person in a trade union, because they cannot disseminate it further if it is classified.

A trade union representative also believed that it was challenging to conduct tripartite cooperation on security due to the transparency limitations and he described that less transparency made the interaction between the parties more difficult. Less access to information also resulted in a lack of legitimacy of security measures implemented on offshore and onshore facilities:

> What you can risk is that those who work in the companies will be negative about some of the necessary security measures, because they do not know why this is necessary. If there is good information about the need for those measures, within the framework of security classification and what must be kept out of the public eye, it will improve the situation.

A high degree of openness is an institutionalized pattern of action and part of the interaction practice in the Safety Forum, and most of the informants expressed expectations of openness. Nevertheless, they accepted that security entailed limitations in openness and information sharing. The representatives trusted the authorities' intentions and actions and accepted that the industry was under a security threat. This breach with the collective expectations was problematized to a surprisingly limited extent. Through the interviews, several different arguments for acceptance of change emerged, with trust as a main factor. Trust between the parties is crucial for the tripartite cooperation to work (Ministry of Labour, 2013), and trust appeared to have acted as a "buffer" and provided institutional flexibility. Through the interviews, we saw expressions of a high degree of trust: the parties trusted each other and had confidence in the network and the institutional environment of which they were a part. Informants from both the employee and employer sides acknowledge that there are limits to openness about security and expressed trust in the authorities' intention to share. Instead of suspicion, the trust-based system provided stability for the rules of cooperation.

6 Discussion

6.1 Trust under Pressure: How Geopolitics Is Reshaping Norway's Tripartite Model

The tripartite model faces significant challenges in balancing regulatory compliance with values-based leadership, particularly in the context of uncertainties like geopolitical instability. Over time, adjustments have been made to balance the need for information sharing with the imperative to protect sensitive data. New patterns of information sharing have emerged, with some exchanges now occurring outside the Safety Forum, signaling shifts in established practices. The heightened focus on security has also impacted trust, introducing elements of distrust that contrast with the Safety Forum's traditional trust-based approach. Nevertheless, trust has been partially preserved through the development of new collaborative frameworks. Securitization has emerged as a key driver of change, placing pressure on the institutionalized model of collaboration and prompting adaptations in interaction practices.

This process of re-institutionalization illustrates how Norway's tripartite cooperation model balances stability with adaptability in the face of evolving challenges. It requires a reconciliation of traditional principles such as openness, trust, and information sharing with new demands for confidentiality, protection, and security. While the model demonstrates resilience and flexibility, tensions have arisen as security requirements conflict with core principles of engagement. If not carefully managed, these tensions could undermine the collaborative model over time.

Changes in the security landscape have affected the interaction, responsibility, and collaboration axis between the parties in working life, evident when representatives of the parties meet in a tripartite context. In the Safety Forum, members on the employer side no longer only represent employers with a classic Health, Safety and Environment (HSE) or societal security responsibility, but to a greater extent also a security policy responsibility. A natural consequence is also that the axis between the employer and the authorities is given greater emphasis, including when it comes to information sharing.

This means that the contrasts between safety and security are not merely about practical or organizational conflicts tied to issues of safety, economy, and managerial responsibility, but also involve questions of values. Employers feel an increased sense of responsibility and obligation due to geopolitical developments. At the same time, labor organizations bear responsibility for their members' safety and security but also carry a political and ideological obligation to consider the nation's best interests. This creates a situation where the parties are compelled into roles as security actors, with value-based questions and their moral responsibility for national security becoming more prominent. There appear to be at least three key motivations for unions seeking inclusion in this context. First, they are likely driven by a genuine concern for the well-being and safety of their members. Second, unions may wish to contribute to national security efforts. Third, they may aim to preserve their existing influence in matters of safety and ensure that this influence extends into the evolving security landscape.

It is this approach to value questions that can be described as a form of value-based re-institutionalization of the tripartite model. Particularly interesting are the positions of labor unions, as they—together with employers and authorities—assume a clear responsibility and moral obligation to contribute to safety and security in a new geopolitical context. In this setting, they collaborate with their "opponents" to a greater extent than before. The geopolitical situation thus leads to increased conformity in the workplace, resulting in changes to the interaction and power dynamics between the parties.

The companies and their employees have become security actors (Hansen & Fiskvik, 2025; Petersen, 2020). The goal of the authorities' information sharing is to create legitimacy for security measures, but also to enable the recipients to be co-creators of security—that is, information is shared to engage the workers in identifying threats. Openness is used as a deliberate means to "mobilize and engage" the workers to share information, be prepared, and be better able to recover. In this study, we see that the informants express a desire to gain access to information in order to be able to share back.

6.2 Consequences for Workers

Many Safety Forum members represent workers directly involved in offshore operations. The changes observed in Safety Forum activities thus provide a window upon what the increased focus on security means for all workers. Based on our empirical observations and broader work, we will highlight three major challenges for workers.

First, workers may face a new kind of moral pressure as security moves to the foreground. Decisions are faster and more centralized, which weakens the participatory norms they are used to in safety. That creates an "informed-but-uninvolved" dilemma: they are asked to comply with intrusive controls (monitoring, access restrictions, rapid role changes) without the context needed to judge proportionality or raise reasoned objections. Secrecy also blurs consent. People cannot meaningfully consent to surveillance they cannot fully understand, yet refusing can mark them as non-compliant. Add security clearances and you get a two-tier workforce where some can know and contest decisions while others must live with them, raising fairness and trust issues on the shop floor.

A second challenge is ethical conflict and responsibility drift. Workers are pulled between defending national interests and defending workers' rights: privacy, due process, equitable treatment—often in the same shift. If a security measure degrades safety (e.g., discourages weak-signal reporting or slows emergency access), who owns that trade-off? When lines of authority stretch into the state's security apparatus, accountability can migrate upward while blame sticks downward. That heightens the ethical risk of scapegoating front-line staff after incidents they had little power to shape, and it chills speaking up—especially when disclosures might touch classified information.

Third, there is the everyday ethics of burden and recognition. Security measures can add cognitive load, travel and roster disruption, background checks that spill into private life, and the stress of being a potential target—all unevenly distributed across roles and contractors. If those burdens are not offset (training, hazard pay, time limits, clear review dates, and credible channels to challenge harms), workers experience security as something done to them rather than with them. Sustaining legitimacy, then, hinges on proportionate secrecy, explainable justifications at the lowest possible classification, and real avenues for worker voice—even in the gray zones where full transparency isn't possible.

7 Conclusions

The Norwegian model emphasizes functional requirements and clear accountability, aiming not just for compliance, but for ethically defensible practices that protect people, the environment, and society. Safety and security are treated as ethical obligations grounded in responsibility, transparency, and fair risk distribution—reflecting core values like trust, equality, and democratic legitimacy.

In a geopolitical landscape characterized by hybrid threats, the significance of Norwegian fossil energy supplies to European security is increasing. In this chapter, we have explored the ethical challenges and compromises faced by the petroleum sector and its workforce following the sabotage of the Nord Stream gas pipelines. We examined how changes in the threat environment and increased focus on security have influenced the collaborative practices between authorities, employers, and employees within the institutionalized tripartite model of the petroleum industry.

Democratic labor relations are fundamentally built on transparency, participation, and trust, particularly in cooperative models where employees, employers, and governments work together to ensure fair working conditions and workplace safety. The tripartite model remains a cornerstone of labor relations in Nordic countries, playing a vital role in regulating working conditions and industrial safety. By integrating perspectives from civil society and working life, the model reinforces ethical values and democratic principles, striving to institutionalize social trust, inclusivity, and shared decision making. This approach ensures that safety and security regulations remain aligned with broader societal values. However, while there is broad consensus on the need for enhanced security measures, we have showed how tensions have emerged regarding transparency, as security concerns often necessitate confidentiality. Openness and trust remain a stabilizing force in tripartite institutions, yet ongoing geopolitical uncertainties continue to challenge the model. Thus, the shifting roles and responsibilities of industry following the new geopolitical situation have direct implications for working life and institutions that structure the relationship between the parties in the tripartite model.

The new framework conditions may impose further challenges on cooperation and dialogue in the sector, reminiscent of the dynamics observed during the initial crisis phase following the Nord Stream sabotage. There is, however, uncertainty concerning the future trajectories and implications for tripartite cooperation, as well as the overall impact of geopolitical tensions on the sector. Nevertheless, this study distinctly indicates that the new framework conditions present ethical dilemmas that require resolution, with various stakeholders holding differing interpretations and viewpoints.

As security concerns move to the forefront, workers face new moral pressures and weakened participatory norms, creating an "informed-but-uninvolved" dilemma where they must comply with intrusive controls without context or meaningful consent. Secrecy and security clearances risk dividing the workforce and eroding trust, while responsibility for trade-offs between safety, privacy, and national security often drifts upward, leaving frontline staff exposed to blame. There is reason to believe that sustaining legitimacy in this context depends on proportionate secrecy, transparent justifications, and credible avenues for workers' to be heard, ensuring security is experienced *with* workers, not *against* them.

References

Boute, A. (2022). Weaponizing energy: Energy, trade, and investment law in the new geopolitical reality. *American Journal of International Law, 116*(4), 740–751. https://doi.org/10.1017/ajil.2022.53

Engen, O. A., Lindøe, P., & Hansen, K. (2017). Power, trust and robustness—The politicization of HSE in the Norwegian petroleum regime. *Policy and Practice in Health and Safety, 15*(2), 145–159. https://doi.org/10.1080/14773996.2017.1318485

Engen, O. A., Lindøe, P. H., & Braut, G. S. (2023). Coping with different system logics of standardization in regulatory regimes. Norwegian offshore experience. *Safety Science, 161*, 106079. https://doi.org/10.1016/j.ssci.2023.106079

Equinor. (2025). *Two COSL rigs awarded drilling contracts.* https://www.equinor.com/news/20230831-rig-contracts

Ericson, R. V., & Doyle, A. (2003). *Risk and morality* (Vol. 31). University of Toronto Press.

Forseth, U., & Rosness, R. (2021). Paradoxes of power: Dialogue as a regulatory strategy in the Norwegian oil and gas industry. *Safety Science, 139*, 105120. https://doi.org/10.1016/j.ssci.2020.105120

Hansen, S. T., & Antonsen, S. (2024). Taking connectedness seriously. A research agenda for holistic safety and security risk governance. *Safety Science, 173*, 106436. https://doi.org/10.1016/j.ssci.2024.106436

Hansen, S. T., & Fiskvik, J. (2025). Responding to a new geopolitical reality. NATO and EU strategies of 'whole-of-society' and 'resilience' and implications for corporate actors. *Proceedings of the 35th ESREL & 33rd SRA-E Conference*. pp. 2116–2123. https://rpsonline.com.sg/proceedings/esrel-sra-e2025/pdf/ESREL-SRA-E2025-P1889.pdf

Hansen, S. T., & Moe, E. (2025). Klima og energipolitikk: Suverenitet, legitimitet og sårbarhet. In Ø. Svendsen & K. Haugevik (Eds.), *Dilemmaer i norsk utenrikspolitikk*. Cappelen Damm Akademisk.

Harris, C. E. (2008). The good engineer: Giving virtue its due in engineering ethics. *Science and Engineering Ethics, 14*, 153–164. https://doi.org/10.1007/s11948-008-9068-3

Jore, S. H. (2019). The conceptual and scientific demarcation of security in contrast to safety. *European Journal for Security Research, 4*(1), 157–174. https://doi.org/10.1007/s41125-017-0021-9

Jore, S. H. (2023). On security and safety. *Risk Discourse and Responsibility, 336*, 232.

Maslen, S., Hayes, J., Holdsworth, S., & Sandri, O. (2021). When ethics is a technical matter: Engineers' strategic appeal to ethical considerations in advocating for system integrity. *Science and Engineering Ethics, 27*(4), 46. https://doi.org/10.1007/s11948-021-00324-7

Midttun, Ø. (2024). *Fra sikkerhetspolitikk til samhandlingspraksis: Hvordan påvirker endringer i trusselbildet trepartssamarbeidet i petroleumsvirksomheten?* Universitetet i Stavanger.

Ministry of Labour. (2013). *Tilsynsstrategi og HMS-regelverk i norsk petroleumsvirksomhet*. Arbeidsdepartementet. https://www.regjeringen.no/globalassets/upload/ad/publikasjoner/rapporter/2013/utvalgsrapport_hms_regelverk.pdf?id=2159644

Moe, E., & Hansen, S. T. (2025). The ocean and the geopolitics of energy: A question of critical infrastructure? In T. Sattich, M. Grasso, E. Moe, M. Schreurs, & S. Yan (Eds.), *De Gruyter handbook of ocean governance and maritime affairs* (pp. 379–394). De Gruyter.

Norwegian Intelligence Service. (2024). *A changed security situation*. Etterretningstjenesten.

Norwegian Ocean Industry Authority. (2022). Observasjoner av uidentifiserte droner/luftfartøy offshore. Retrieved from https://www.havtil.no/utforskfagstoff/fagstoff/fagartikler/2022/observasjoner-av-uidentifiserte-dronerluftfartoy offshore/.

Norwegian Ocean Industry Authority. (2023). *Security against deliberate attacks*. http://www.havtil.no/en/explore-technical-subjects2/technical-competence/news/2023/topic-security-against-deliberate-attacks/

Norwegian Police Security Service. (2024). *National Threat Assessment 2024*. https://www.pst.no/globalassets/2024/nasjonal-trusselvurdering-2024/nasjonal-trusselvurdering-2024_engelsk_web_.pdf

Petersen, K. L. (2020). Three concepts of intelligence communication: Awareness, advice or co-production? In *Intelligence on the frontier between state and civil society* (pp. 7–18). Routledge.

Petersen, K. L. (2023). Ukraine og enden på den private sektors uskyld. *Politica, 55*(1), 74–85.

Petroleum Safety Authority Norway (PSA). (2017). *Sikkerhet og ansvar. Forstå det norske regimet*. https://www.ptil.no/contentassets/0079bf5eb8824beb969fd0f217f395b7/sikkerhet-og-ansvar.pdf.

Petroleum Safety Authority Norway (PSA). (2022). *Sikringstiltak må fungere*. https://www.ptil.no/fagstoff/utforsk-fagstoff/fagartikler/2022/sikrings%2D%2Dog-beredskapstiltak-ma-fungere/

Piètre-Cambacédès, L., & Chaudet, C. (2010). The SEMA referential framework: Avoiding ambiguities in the terms "security" and "safety". *International Journal of Critical Infrastructure Protection, 3*(2), 55–66. https://doi.org/10.1016/j.ijcip.2010.06.003

RAND Corporation. (2024). *Evolving Norway's role in the NATO Alliance*. https://www.rand.org/pubs/commentary/2024/07/evolving-norways-role-in-the-nato-alliance.html

Skare, E., & Jore, S. H. (2024). Hybrid threats in the Norwegian petroleum sector. A new category of risk problems for safety science? *Safety Science, 176*, 106521. https://doi.org/10.1016/j.ssci.2024.106521

Storesund, R., Jore, S. H., & Antonsen, S. (2024). Safety science in a hyperconnected world. *Safety Science, 177*, 106589.

Van de Poel, I. (2011). The relation between forward-looking and backward-looking responsibility. In *Moral responsibility: Beyond free will and determinism* (pp. 37–52). Springer. https://doi.org/10.1007/978-94-007-1878-4_3

CHAPTER 10

Ethical Licensing in Practice: The Case of CRISPR Gene Editing Technology

Alison McLennan and Sarah Maslen

1 Introduction

CRISPR Babies

In November 2018, news broke that scientists in China had used CRISPR-Cas 9 genome editing to make heritable changes to germline cells in human embryos. These were then implanted into women, resulting in the birth of two human baby girls with edited genes (Cyranoski & Ledford, 2018). This was the first use of genome editing in humans (Feeney et al., 2021). The scientists claimed to

A. McLennan (✉)
Canberra Law School, Faculty of Business, Government and Law, University of Canberra, Canberra, ACT, Australia
e-mail: alison.mclennan@canberra.edu.au

S. Maslen
School of Property, Construction and Project Management, RMIT University, Melbourne, VIC, Australia
e-mail: sarah.maslen@rmit.edu.au

J. Hayes, S. Maslen (eds.), *Ethics in the doing of technoscience*,
https://doi.org/10.1007/978-3-032-20002-0_10

have edited the genes to reduce HIV susceptibility in the resulting babies (see Greely, 2019). The Chinese scientists appeared to have performed their research without submitting to ethical or legal oversight (Feeney, 2019; Feeney et al., 2021).

This was met with condemnation by scientists and bioethicists around the world, including the pioneers of CRISPR-Cas 9 technology. It was seen as "premature" and "dangerous", exposing children to risks without addressing a medical need (Cyranoski & Ledford, 2018). Scientists at the International Summit on Human Genome editing where this research was presented quickly released a statement condemning this "deeply disturbing" experiment as irresponsible and contrary to international norms and ethical standards, alongside scientific criticisms of the protocol (Organizing Committee of the Second International Summit on Human Genome Editing, 2018). They raised concern about the risks and uncertainty associated with heritable human genome editing. They called for a "broad public dialogue", the development of tools to assess risks and "common regulatory standards". Broad Institute Scientist David Liu expressed a similar sentiment:

> The recently reported use of CRISPR nuclease to edit CCR5 in human embryos … represents a serious breach of ethics that I hope will serve as a wake-up call for the community. Foremost, that edited human babies were generated without the full engagement of independent scientific and ethics experts, relevant regulatory institutions, and governing bodies is appalling. In addition, the unmet medical need motivating the work is unclear at best.

This incident focused attention on the regulation of CRISPR and raised questions about whether it was sufficient given the rapidly developing science. Governments and scientists around the world are now actively considering these questions. He Jiankui and colleagues were sentenced to three years in jail. They were found to have broken laws, avoided ethical review by forging documents, and misled doctors about the nature of the embryos involved (Normile, 2019).

CRISPR-Cas 9 is a genome editing system in which scientists use an enzyme called Cas 9 along with a guide RNA (or ribonucleic acid) molecule to make precise, targeted changes to DNA (deoxyribonucleic acid) sequences (Broad Institute, n.d.-a; Ledford, 2015). CRISPR-Cas 9 is the original CRISPR technology, and CRISPR techniques using other enzymes are now becoming available. We will use "CRISPR-Cas 9" to refer to this original technology and "CRISPR" to refer to the suite of genome editing systems now in use. CRISPR techniques allow researchers to "permanently modify genes in living cells and organisms and, in the future, may make it possible to correct mutations at precise locations in the human genome in order to treat genetic causes of disease" (Broad Institute, n.d.-a). The development of this method for modifying genomes has greatly advanced the field of synthetic biology, hailed as the "Holy Grail" of biotechnology due to its applicability to a wide range of organisms (Scheinerman & Sherkow, 2021, p. 2). However, it has also come with some challenges for researchers who wish to work with the technique, which center on the ethical questions that the technology raises.

The "CRISPR babies" incident and the ensuing moral panic brought the power of CRISPR-Cas 9 and its ethical issues sharply into focus, not only for scientists but also for governments and the broader public. In 2021, genetic enhancement of children even became the subject of a dystopian novel, *Klara and the Sun* (Ishiguro, 2022). Research since the CRISPR babies incident has highlighted public concern about the technology's potential uses in humans (Scheinerman & Sherkow, 2021). This debate has distinguished between changes to two types of cells: somatic cells and germ cells (see Greely, 2019). Changes made to somatic cells are not heritable, whereas changes made to germ cells (germline editing) are heritable. He Jiankui claimed to have carried out germline editing on cells in human embryos (see Greely, 2019). There is "rough consensus" that germline editing of humans should be prohibited, at least for reproductive purposes (Feeney et al., 2021, p. 3). Among the public, the significant concerns about CRISPR gene editing technology relate to nonclinical uses such as human enhancement (Nicol et al., 2022, p. 23) and access to CRISPR technology's benefits (Nicol et al., 2022, p. 59; Scheinerman & Sherkow, 2021). The public is also concerned about appropriate regulation of this technology (Nicol et al., 2022, pp. 20–22).

Generally, scientific research is subject to ethical and safety oversight imposed by governments via legislation. There are institutional ethical and safety approval checks performed under a legislative framework, and for some research external approvals are also required. For example, in Australia where we conducted our research, some scientific research requires approval from the independent Gene Technology Regulator. These mechanisms can be described as "public governance" of the technology, as they are underpinned by governments' legislative authority. He Jiankui and collaborators flouted these usual controls on how new technologies are used. This was a key reason for the condemnation that followed (see Organizing Committee of the Second International Summit on Human Genome Editing, 2018; Feeney et al., 2021, p. 3).

The use of technology in scientific research is also subject to further legal control through the patent system. Patents give an inventor a time-limited exclusive right to exploit their invention and to authorize others to exploit it (in Australia, see *Patents Act 1990* (Cth) s 13). This means that where technology is patented in a particular jurisdiction, users in that jurisdiction would need to obtain a license from the patent holder to use that technology legally. In practice, patent holders usually hold patent rights in more than one country to maximize the market for their invention meaning their patent rights are international in scope (see IP Australia n.d., Patent Cooperation Treaty, 1970). In this way, patents are "legal instruments" that "operate as a right to exclude others from using, selling ... particular inventions" (Scheinerman & Sherkow, 2021, p. 3).

In addition to being powerful and widely used, CRISPR-Cas 9 is commercially valuable (Scheinerman & Sherkow, 2021, p. 2). The patent landscape surrounding CRISPR-Cas 9 is complicated and contested (Sherkow, 2025). The foundational patents on the original CRISPR-Cas 9 technology are held by multiple entities including the Broad Institute (a joint venture between the Massachusetts Institute of Technology and Harvard University) and the University of California, Berkeley. There is much commercial research utilizing CRISPR-Cas 9 that requires access to the foundational technology and would be vulnerable to litigation if licenses were not obtained giving the patent holders a powerful position in imposing license conditions (Feeney et al., 2021, p. 4).

While the economic drivers of securing patents are key, patent holders can also act in ways that do not immediately serve to maximize their profits but are rather grounded in other professional values. Some entities that own CRISPR-Cas 9 patents use this property right to decide how the technology is used. Patent holders thus seek to use their patents "as a tool to foster ethical conduct" (Nicol et al., p. 62), giving licenses only for research that they deem ethical. This is known as "ethical licensing restrictions" or "governance by patent" (Scheinerman & Sherkow, 2021; Sherkow et al., 2021). For some research, licenses can be required from multiple patent holders, who may each make judgments over the ethics of proposed uses ahead of granting the license. This chapter focuses on the practice of patent holders acting as decision makers on ethical use of CRISPR-Cas 9. These practices raise questions about whether ethical licensing is an appropriate use of the patent system. We look at this issue from the perspective of Australian scientists seeking licenses, offering rare empirical evidence of ethical licensing and how it affects the doing of the scientific research in practice. We argue that governance by patent is ineffective as a regulatory tool in practice, as well as being an undemocratic, unintended use of the patent system. We call for greater transparency and oversight, as well as clear boundaries for scientists.

2 Who Decides "Ethical Use"?

At the international level, regulation of the use of CRISPR is a "partial, ineffective patchwork" (Feeney et al., 2021, p. 2). The "common regulatory standards" that scientists advocated for in 2018 have not materialized. Instead, there are diverging approaches across jurisdictions (Nordberg et al., 2020). In this context, some CRISPR-Cas 9 patent holders have used their position as patent holder to impose pseudo regulation. One of the key players, the Broad Institute, is currently allowing "academic and non-profit research" using its CRISPR-Cas 9 technology to occur without requiring researchers to seek a license (Broad Institute, n.d.-c). This approach aims to facilitate research in the field (Broad Institute, n.d.-c). However, research with a commercial goal in mind requires a license in order to use the patented technology. This means that the Broad Institute can be selective about how its patents will be licensed. As we will see, it chooses not to issue licenses for uses it considers ethically problematic.

Questions about what use of a technology is ethical are not always as black and white as in the CRISPR babies case. Some applications of CRISPR have complex ethical considerations associated with them, and a diversity of views may be present in the scientific community and the public. By imposing ethical licensing conditions, patent holders take on the role of ethical arbiter. Ethical gatekeeping by patent holders is a practice with significant real-world impact. As commentators have pointed out, "Policies concerning patent licensing have the potential to ultimately affect access, distribution, and conditions on use of particular technologies" (Scheinerman & Sherkow, 2021, p. 3). In this way, restrictions on licensing actually "have potential to speed or hinder technological development, to place it in the hands of a select few or many" (Scheinerman & Sherkow, 2021, p. 3). Patent licensing decisions can also affect the course of research and ultimately the cost of therapies (Scheinerman & Sherkow, 2021, pp. 6–7).

There are two main positions on the question of whether ethical licensing is a desirable regulatory approach. On the one hand, some argue that patent holders have the right to manage their inventions and ensure that they are used responsibly. The argument is essentially that patents are designed to allow the patent holder to control the use of its technology (Scheinerman & Sherkow, 2021, p. 2). As Scheinerman and Sherkow explain, "this right to license brings with it a right to establish licensing conditions governing specific uses or development restrictions over a particular technology, i.e. barring licensees from engaging in particular veins of research or development" (Scheinerman & Sherkow, 2021, p. 5).

A further argument in support of ethical licensing is that patent holders are well placed to make decisions about social and ethical issues surrounding the technology because of their scientific technical expertise (Scheinerman & Sherkow, 2021, p. 5; Guerrini et al., 2017). They can "identify controversial applications" (Guerrini et al., 2017), understand the positive and negative uses, and decide "what is possible, realistic, genuinely dangerous" (Feeney et al., 2021, p. 5). Some scholars observe that regulation by patent holders could still be receptive to "participatory democracy" and that governments could intervene if there were license agreements that were contrary to public policy (Scheinerman & Sherkow, 2021, p. 5). Some have also argued that patents should be part of the regulatory mix because the existing governance mechanisms and bodies that implement these are "ill equipped to address complex societal and value-based concerns in an increasingly privatized world" (Parthasarathy, 2018, p. 288). Pragmatically, some scholars argue that ethical licensing is better than no regulation, given that there is no current global consensus and that consistent international regulation will develop slowly (Feeney et al., 2021). As patent holders can seek rights in multiple jurisdictions, ethical governance by patent could occur in multiple jurisdictions (Feeney, 2019).

In this vein, ethical licensing can also be seen as advantageous for reasons of efficiency (Guerrini et al., 2017). Only one party's view is relevant, so it can be implemented much more quickly than traditional regulation, which requires consultation with multiple stakeholders (Feeney et al., 2021, p. 5; Guerrini et al., 2017). Ethical licensing can also be seen as more flexible than traditional regulation as decision-making can quickly be changed in response to new information becoming available (Feeney et al., 2021, p. 5). This makes it a much more agile regulatory approach than amendment to legislation considering new scientific data (or changes in public perceptions of the technology and its acceptable use). Lastly, license restrictions imposed by patent holders can be enforced through the courts (Feeney et al., 2021, p. 5; Guerrini et al., 2017). Some have suggested that this voluntary approach taken by patent holders could even result in "increased faith in scientific self-regulation" (Guerrini et al., 2017).

Conversely, some commentators contest whether patent holders should have so much power over scientific and medical developments. They consider decisions about ethical use should be made in a broader context rather than by private enterprises. In short, while ethical licensing may not be illegal, it is a misuse of the patent system. The key criticism here is that ethical licensing is undemocratic (e.g., Scheinerman & Sherkow, 2021). Feeney and colleagues note, "It is a form of privatized morality—without discussion, debate, public involvement and democratic accountability—that forecloses ethical decision-making on the technology with wide societal impact". This means that as a regulatory tool, ethical licensing lacks "democratic legitimacy" (Feeney et al., 2021, p. 2).

Concerns around access to the benefits of CRISPR technologies raise questions of fairness and justice. Scheinerman and Sherkow explain that "CRISPR's power to treat if not cure a great many unevenly distributed genetic diseases means that the distributional choices made by licensing governance is likely to have an outsized effect on the polity" (Scheinerman & Sherkow, 2021, p. 7). Accordingly, it is seen as important to include the public in policymaking about patent licensing (Scheinerman & Sherkow, 2021, p. 9). Commentators have expressed concern that ethical licensing thus gives too much discretion to patent holders and that "leaving the determination of what is 'socially responsible' to the sole discretion of the patentee, ethical licensing through private governance raises procedural justice concerns" (de Graeff et al., 2018, p. 58). Self-regulation by patent holders has been seen as "insufficient from a social justice perspective" to regulate the use of CRISPR due to the way in which decisions would be made by only a small number of powerful commercial entities (Feeney, 2019, p. 229).

There is also opposition to ethical licensing on pragmatic grounds; it is ineffective as a regulatory tool. These concerns center on the voluntary nature of the restrictions on gene editing through patent licensing (Feeney et al., 2021, p. 1). Some argue that that this should not be seen as a stable, sustainable form of regulation because we do not know how long the current position of the patent holders will last (Feeney et al., 2021, p. 6). While some suggest that maybe patent holders would be receptive to public attitudes toward gene editing (Scheinerman & Sherkow, 2021, p. 2), others stress that profitability is a competing consideration (Feeney, 2019, p. 229; Feeney et al., 2021, p. 6). Further, the private governance mechanism is likely to be inconsistent between private actors and between jurisdictions (Feeney, 2019, p. 229; Feeney et al., 2021). This creates its own risks. As Feeney et al. (2021) put it: "the greater the divergence in international governance … the greater is the risk of unscrupulous actors, companies or indeed states moving genome editing operations to other locations where there are no prohibitions or other restrictions" (p. 2).

As an alternative to the choice between self-regulation or regulation by government, a third option has been proposed involving regulation through the patent system, led by government (Parthasarathy, 2018). In this model, an advisory body would guide decisions about controlling CRISPR patents. This is said to have advantages in terms of "transparency and legitimacy" as well as "speed and stability" (Feeney et al., 2021, p. 2). However, concerns about democratic legitimacy may still arise.

While debates over the appropriateness of ethical licensing are extensive, there is a lack of empirical examination of how it operates in practice. This chapter makes a contribution to this research gap.

3 Our Research

Our research was designed to investigate how the intellectual property (IP) law and policy environment affects Australia's synthetic biology industry. This project adopted a mixed methods approach including interviews, surveys, and analysis of laws and other documents to explore "law in the real world" as a basis for making recommendations for law or legal policy (Bell, 2016). In this chapter, we address the issues scientists encounter in ensuring freedom to operate and securing appropriate licenses as they were explored via interview. We focus on two scientists who gave accounts of their experience of ethical gatekeeping by patent holders.

Interviews took place in 2021–2022 and were conducted by the first author, who is both a legal scholar and has an undergraduate background in biology. Due to social distancing requirements, they generally took place over Microsoft Teams, with two interviews conducted in person. The interviews lasted around one hour on average. Interviews were audio recorded for analysis with the consent of participants. A total of 22 semi-structured interviews were conducted as part of this study. Fifteen of the interviews were conducted with scientists working in synthetic biology in Australia. These scientists were working in universities and other research settings such as start-ups and larger commercial enterprises and in various areas including microorganisms, plants, and animals. A further seven interviews were conducted with legal, commercialization, and technology transfer professionals. Human research ethics committee approval was granted for this research (University of Canberra [1705]; CSIRO [052/19]).

Interviews were professionally transcribed. We conducted a thematic analysis (Ezzy, 2013) and generated vignettes. The two authors thematically coded the data independently, and then both authors reviewed and refined the codes and themes for the analysis. In the interviews generally, patents relating to CRISPR-Cas 9 technology were identified as the most significant intellectual property concern for these scientists (McLennan & Maslen, 2023). Our analysis presented in the following was particularly attentive to the scientists' experiences with accessing CRISPR-Cas 9 patents, and their decisions as to how to manage it. Participants are referred to by a single letter assigned chronologically to maintain anonymity. We contextualize these experiences via a content analysis of the public statements of the Broad Institute, its scientists and other scientists and scientific organizations.

4 Findings

4.1 *The Broad Institute on "Responsible Innovation"*

The Broad Institute considers its role to include considering ethical issues associated with the use of its technology:

> We are committed to making these technologies broadly available for research to help ensure that therapeutic development—bringing this

technology to the clinic—has the best chance of success. We also consider important ethical, safety and societal concerns. (Rozen, 2016b)

With this in mind, the Broad Institute aims to take a leadership role in ensuring responsible innovation:

To ensure that scientific innovation benefits human health, scientists must also try to ensure that cutting-edge technologies are used in a socially responsible way. We may place policy restrictions on certain licenses to prohibit uses that would be socially irresponsible based on current scientific knowledge and societal consensus. (Broad Institute, n.d.-b)

Further, the Broad Institute has made clear that it will not license its patented technologies for any clinical use involving human germline editing (Broad Institute, n.d.-c). Scientists at the Broad Institute have issued public statements regarding what they consider to be ethical or unethical use of CRISPR technologies. They are in agreement that human germline editing as in the "CRISPR Babies" case should be prohibited for now (Broad Institute, n.d.-b). Many of the publicly available comments on human germline editing are positioned against the "CRISPR Babies" case. For instance, David Liu, a Broad Institute scientist, said:

I anticipate a future in which the application of genome editing technologies provide therapies or even cures for grievous genetic diseases for which no treatment is known. In somatic cells, which do not become part of future generations, these technologies can be used in a focused manner and with the consent of the patient. Embryo editing, in which changes are made to the genome of cells that may include the sperm and egg cells that become future generations, is a very different situation that raises additional critical scientific and ethical issues. There may be a future in which embryo editing is justified following a careful analysis of potential benefits, risks, and ethical issues, but the recently reported work does not meet these standards. Because of the potential of genome editing to provide a transformative benefit to patients, it is essential that we progress in a thoughtful and coordinated manner. Failure to do so risks both harming patients, and delaying or denying potential therapies to many that urgently need them.

One of the developers of CRISPR-Cas 9, Feng Zhang, responded in this way:

> Given the current early state of genome editing technology, I'm in favor of a moratorium on implantation of edited embryos ... until we have come up with a thoughtful set of safety requirements first. Not only do I see this as risky, but I am also deeply concerned about the lack of transparency surrounding this trial. All medical advances, gene editing or otherwise and particularly those that impact vulnerable populations, should be cautiously and thoughtfully tested, discussed openly with patients, physicians, scientists, and other community members, and implemented in an equitable way.

This statement points to multiple ethical concerns, including safety, transparency, and use for vulnerable populations.

Another Broad Institute scientist Eric Ladner identified four key issues: precision of the technology, medical need, the consent of future generations, and finally morality. He described the moral issues as follows:

> What's right and wrong and how we ought to live as a society. Although scientists may be reluctant to debate ethics, we have a responsibility to do so and insights to offer. How would routine genome editing change our world? Would we come to regard our children as manufactured products? Would marketers shape genetic fashions? Would the "best" genomes go to the most privileged? If we cross this threshold, it's hard to see how we could ever return. ... A ban could always be reversed if we become technically proficient, scientifically knowledgeable, and morally wise enough and if we can make a compelling case. But authorizing scientists to make permanent changes to the DNA of our species is a decision that should require broad societal understanding and consent. It has been only about a decade since we first read the human genome. We should exercise great caution before we begin to rewrite it.

While each of the scientists does not close the door on human germline editing in the future, on ethical grounds, they are in agreement that it is not appropriate at this stage.

While some would make a case that using gene technology on animals also involves ethical choices, we did not find public statements regarding the ethics of using CRISPR-Cas 9 in animal research. We found no articulation of principles on which license requests for this work would be assessed. There are however examples of licenses granted by the Broad Institute to other institutions for use of CRISPR to make "model organisms" (e.g., The Jackson Laboratory, 2025). Model organisms are organisms, such as mice, which are used to study conditions which affect humans. For example, scientists may modify the genome of a mouse to generate cancer. Such mice can then be used in research into therapeutics for treating the disease, with the goal of eventually translating this into human therapeutic applications.

When it comes to agricultural research, the Broad Institute notes that there are already regulatory agencies with oversight of many of its concerns (Rozen, 2016a). Nonetheless, they state: "The Broad feels it is important to include explicit restrictions in the technology licenses as well. We wanted to share our thinking with others who may be considering licensing of related technologies". The licensing exclusions are work involving "gene drives", "sterile seeds", and certain tobacco research (Rozen, 2016a).

Gene drives "increase the rate at which certain genes are inherited by the offspring of a sexually reproducing organism" (Australian Government, 2024). They are controversial because of potential risks to the environment. The Broad Institute has stated that gene drives are "a way to rapidly spread a new gene throughout an entire species in nature. This approach might be used to block the transmission of malaria by mosquitoes, but has the potential to disrupt ecosystems" (Rozen, 2016a).

"Sterile seeds" refers to plants which are genetically modified so that they will not produce fertile seed. This technology was highly controversial when developed by biotech giants such as Monsanto in the 1990s because of the potential impact on farmers, to the degree that they did not proceed to commercialize this technology (e.g., Niiler, 1999). The Broad Institute states that this technology would allow seed companies "to force farmers to purchase seed each year" raising ethical concerns regarding fairness (Rozen, 2016a). The Institute agrees with the United Nations Convention on Biological Diversity recommendation that there should be a global moratorium on the development of sterile seed technology (Rozen, 2016a).

Certain tobacco research is considered ethically problematic (Rozen, 2016a). These specific prohibitions do not reflect a broad ethical concern with the use of CRISPR in plant research using the tobacco plant, but specific concerns about the harmful effects of tobacco smoking on human health.

The Broad Institute explains that "we would be concerned about efforts to use genome editing to increase tobacco usage. However, we know that tobacco plants can be used as an important model organism for basic research and can be used as a way to manufacture therapeutic proteins" (Rozen, 2016a). These types of tobacco research are not prohibited.

What is lacking is an articulation of what the Broad Institute regards as ethical or "socially responsible". That is, the Broad has not publicized general principles which can be used in practice to assess the ethics of proposed CRISPR usage. Indeed, many public statements come from individual scientists, rather than the Institute.

5 Doing Scientific Research in the Shadow of Ethical Licensing

5.1 Making Knockout Animals

Participant V is a team leader in a lab that conducts genetic editing of animals. V understands the practice of synthetic biology as organizing genetic material "in a way that couldn't happen in nature". V's work involves making so-called knockout animals. A genetic knockout "refers to the use of genetic engineering to inactivate or remove one or more specific genes from an organism. Scientists create knockout organisms to study the impact of removing a gene from an organism, which often allows them to then learn something about that gene's function" (National Human Genome Research Institute, 2025).

V was conscious of concerns about gene editing among the general public, which they partly attributed to a lack of knowledge. V spoke about how members of the public may be reluctant to use vaccines made with synthetic biology tools, not realizing that many existing vaccines are genetically modified. They noted that people are more open to medical technologies than modified foods:

> So if it's for medicine, I think people's acceptance of it is a lot different than if it was for eating, like if I said I was going to make some crazy chicken that you were going to eat, people would be less accepting of it.

V does not share these public concerns, focusing instead on the good that comes from applications of the technology. They explain that they can take a gene from one animal and put it into another, which can have a range of "useful, beneficial" outcomes including improving disease resilience in the animal.

V uses a variety of technologies and techniques in the laboratory, from glass capillaries to inject embryos with genetic material to gene editing methods like CRISPR-Cas 9. Because CRISPR-Cas 9 can be used without a license in fundamental or "basic" research, V's team can use the patented technology for research that is not commercial, including the creation of knockout animals. For CRISPR-Cas 9, V identified two complicating factors for commercial research. The first is that there are multiple companies which hold patents on this technology. The second is that these patent holders have positions on what they consider ethical use of the technology that can be both unclear and not aligned with V's own view about ethical practice. When describing this, V noted that the patent holders' approach is more (or less) clear cut depending on the research domain:

> I think they're pretty much you can use it for plants. You can use it for medical things, you can use it in cells, you can use it like lots of things are really clear cut. ... For some things like they're never going to give you a license to do something in human embryos ... but they haven't come out and said yes or no for animals. So you don't know. ... So yeah, makes it just makes things a bit more muddy.

While in this summary of the situation, the ethical position of the Broad Institute on using CRISPR-Cas 9 in animal applications is unclear, V explained that the potential for roadblocks is itself enough to prevent the research from going ahead due to concerns by funding bodies:

> We've had issues where we've gone to have contracts with companies, where if we're using CRISPR, we've hit issues because if we're going to use that technique there's no clear IP, so you sort of hit roadblocks there.

V continued with the research context of the knockout animal example:

> If we were going to make a [modified animal] for someone external, which you know we've got the capability of doing that. But if we wanted to use CRISPR to generate, say that [animal] ... there's no clear way that we can get a license, like for research, you can do it, but if we wanted to, if someone was paying us to do it as a commercial thing, we can't get the license to use CRISPR, so that's an issue.

When V describes projects that have encountered a roadblock to their commercial partnerships in relation to CRISPR-Cas 9 patents, it has taken place at the funding stage. Commercial funders anticipate CRISPR patent issues, and so scientists know that if they put up proposals that adopt CRISPR then the potential commercial partner may well reject the collaboration as too risky.

In V's view, the "muddiness" does not stem from ethical concerns. Instead, it reflects a lack of consideration of such applications, due to market factors:

> And it's the two different companies that have got the rights that. … It's something that they … haven't really had to consider whereas, you know, modify plants and using it for medical applications, there's plenty of those on the market, so they've concentrated on that. … But they haven't necessarily thought about animal biotech, because I guess it to them it's not a big market at this point in time.

The implication here is that, if the Broad Institute considered its case of animal research closely, it would share its view of this appropriateness. As V noted, its research aims include reducing culling of animals, something that CRISPR techniques can facilitate as described above.

V felt that while ethical licensing of CRISPR-Cas 9 is causing problems, there are other ways to do the work and so it does not change the outcome of their practice: "I think people would know the issues with CRISPR and then they are doing these modifications of that allow them to get around it as well". V mentioned other gene editing tools which have what V described as "clear IP". V said, "There are more enzymes coming out … that are slightly different but again, some of them haven't made clear what their plans for it are". Here V is referring to other gene editing tools, which may be licensed by further patent holders who have not yet stated whether or how they will engage in ethical licensing restrictions.

5.2 Engineering Microorganisms

Participant G is a senior scientist working for a research institution. G considers synthetic biology to be an extension of "fundamental biological concepts … applying engineering principles so that we can create *useful applications … and useful technologies*" (emphasis added). G hopes that by using computations tools and these engineering principles, synthetic biology will allow scientists to "do the sorts of engineering of biology that we used to do … at a larger scale, more accurately and precisely and more quickly". In G's articulation of synthetic biology, there is a sense that this work serves a public good. G's research group is working on engineering microorganisms so that they produce particular chemicals. This requires making many variations to the microbial DNA and analyzing the resulting chemicals to assess their properties. G did not express awareness of any public contestation over the ethics of this work.

Given the nature of their research, G's research organization uses gene editing methods like CRISPR-Cas 9 and other similar tools. Like V, G's understanding is that patents do not create impediments to "basic" research. However, G considers that complications can arise linked to ethical licensing in seeking licenses for research with a commercial outcome.

Where a scientist requires access to CRISPR-Cas 9 as part of their research, G explains that it is unclear who would need to grant a license: "There are several claims on ownership of that technology. … it's not at all clear with whom we have to negotiate a position". Beyond these issues of the complicated patent landscape, G was aware that the Broad Institute, as one of the patent holders, has strong views and actively practices ethical licensing, which can impede the conduct of what they, as scientists, consider to be ethical research: "Broad Institute has some very definite ideas about what the technology should and shouldn't be used for. So negotiating with them can sometimes be quite trying".

G was thus aware that the patent holder's views about what the technology should ethically be used for had become part of the process of accessing the IP. In practice, an organization like the one that G works for could find itself needing to negotiate multiple licenses for different work groups due to practices of ethical licensing:

> Some folk who, for example, want to use CRISPR in animals have a different or are having to, independently, negotiate a position compared with people who want to use CRISPR in plants, even though we're working, effectively, with the same company.

This suggests that the patent holders consider different ethical issues to be posed by the use of CRISPR in these areas.

Due to these difficulties in licensing and navigating the CRISPR-Cas 9 patent landscape, G's research entity is looking at alternatives "where the IP landscape is considerably simpler and more easily understood". While they noted that the search for workarounds has a cost in terms of delaying research work, they did not indicate that the choice to grant or deny a license resulted in a change to their views about the ethics of a particular project, or whether they ultimately pursued it. They gave the following example of such a workaround when faced with difficulty negotiating access:

> We had a product where we were using one particular plasmid in a bacterium to make a protein. And the company that owns the IP for that plasmid was being quite restrictive, difficult to work with. And so we just made our own plasmids. We worked around it basically. … There was a bunch of work that we had to do to be able to work around that—it does delay the development of products when you have to find these workarounds.

In general, G felt that restrictive licensing practices are inappropriate where foundational technologies are concerned. We can infer that they consider it is not the patent holder's role to evaluate ethical use and grant a license (or not) on this basis, or to otherwise restrict access to key technologies in their field. G would like to see "enabling technologies", like CRISPR-Cas 9, made open access, rather than licensed in restrictive ways:

> Where the application is a broad enabling tool for the community, keeping that information as a trade secret can cause you all sorts of problems. Or cause the community all sorts of problems. As they try to work round it. As they try to, you know, negotiate positions and all those other things. I do think that sitting on patents that allow you to control the use of enabling technologies is a real hurdle and causes all sorts of problems in tech development.

G gave the case of polymerase chain reaction (PCR) technology as a counterexample. The wide availability of this technology supported major scientific advances including the development of recombinant DNA technology. G explained: "The benefit of that is enormous. And you can see the benefits adopted in every lab effectively around the world. And so huge advances were made because it was … just available". This suggests that G thinks that patent holders should not be able to restrict access to foundational research tools, due to the importance of those tools for scientific progress.

6 Discussion

Our findings trace the ethical tensions that are apparent in the doing of synthetic biology research in relation to CRISPR-Cas 9. In setting out the reasons for its ethical licensing restrictions, the Broad Institute articulates an ethics of "social responsibility". While they will give licenses for use of the patented technology in some applications such as plant research, it is unclear whether the Broad Institute will give licenses for using the techniques in animals due to the ethical dimensions of genome editing in animals. For the scientists, their ethics is based on developing useful practical applications of the technologies they work with, as well as a commitment to scientific progress.

In considering how the phenomenon of ethical licensing may impact on scientific practice, it is important to note the other forms of ethical oversight which already guide this practice. Ethical restrictions on biological research may be imposed by legislation, which is then implemented by regulators and institutional ethics committees. Usually, a scientist would consider the ethics of their research, and their project would be subject to institutional ethics approval. As a form of ethical oversight, institutional research ethics committees are rigorous. In Australia, ethics committees assess research involving animals in accordance with the detailed *Australian Code for the Care and Use of Animals for Scientific Purposes* (National Health and Medical Research Council, 2013). This framework provides "governing principles" underpinning the ethical approval of such work (National Health and Medical Research Council, 2013, p. 9). Respect for animals is paramount here. Further governing principles, such as only using animals where justified, supporting their well-being, minimizing harm, and maintaining high levels of scientific integrity, all flow from the principle of respect for animals (National Health and Medical Research Council, 2013, p. 9).

Reflecting on the knockout animal case, irrespective of the patent license exemptions for non-commercial research, the scientists involved would need to have the work approved by an institutional animal research ethics committee before undertaking it. This means that the scientists would need to engage with the ethical guidelines around the care and use of animals at the research design stage. They would need to explain why their work should be considered ethical and be allowed to go ahead. This requirement would be unchanged in the commercial application, raising questions about whether an additional layer of ethical oversight in the form of ethical licensing is justified. Indeed, as licenses for patented technologies are usually sought much later in the research process, engagement with the ethical issues (via ethical licensing) is moved much further downstream, with our findings suggesting it is seen as a bureaucratic hurdle only.

One of the common practices captured in the two cases that we detailed is the adoption of workarounds on the part of the scientists. They both said that they and their institutions were aware of the reality of ethical licensing and while this threatened the speed at which they could progress research, they expected to be able to use alternative enabling technologies to do the work. This raises another challenge to ethical licensing as a pseudo regulatory tool. If the scientists are just doing something else instead, ethical licensing is not an effective way to enculturate norms among scientists. The Broad Institute's current policies focus on specific exclusions and do not indicate general ethical principles. Indeed, even the public statements on exclusions are minimal, as is reflected in our reliance on the statements of scientists rather than the Broad Institute in our presentation of their position. This lack of clear information is also reflected in the comments of scientists who stated there was a lack of clarity over what is, and is not, allowed, though they could make some assumptions. This means that the Broad Institute's ethical licensing approach does not offer a framework of principles which scientists could use to engage with ethical issues. We found no evidence that ethical licensing has changed how individual scientists approach ethics in their work.

While the Broad Institute's ethical licensing approach does not appear to act as a guide to scientists' ethics, the fact that it does make the research process more difficult is significant. It adds to the complexities already associated with accessing patented technologies and ensuring regulatory compliance, which from the scientists' perspectives impedes their pursuit of scientific progress and its applications. Previous research has raised concerns about the patent system's unintended consequences, such as restricted access to patented biomedical technologies, the impact on research trajectories and subsequently on accessibility of treatments and testing for patients (Brownsword & Goodwin, 2012, pp. 290–295; McLennan, 2018; McLennan & Maslen, 2023). Well before CRISPR, in the Myriad litigation over the breast-cancer gene patents and genetic tests, scientists, lawyers, and patient groups expressed concerns about the patent holder using their exclusive rights to control access to genetic testing as well as the direction of research in this area (see McLennan, 2018;

McLennan & Maslen, 2023). The patent rights were used to charge monopoly prices for genetic testing, and patients could not access alternative testing, "doubtlessly resulting in late diagnosis, illness, unnecessary surgery and death" (Feeney et al., 2021, p. 5). Such cases perhaps cause some skepticism on the part of scientists over who should control access to foundational technologies, and how to ensure equitable access to beneficial applications.

At present, the approach of the Broad Institute has been different from the restrictive approach taken by Myriad Genetics in the breast cancer gene case. While the Broad Institute is taking steps to make its patented technology accessible for "public benefit", there are no guarantees that this policy will continue in a situation where "profitability, for instance, is an alternative and competing value" (Feeney, 2019, p. 6). Commentators have concerns that the exclusive licensing model adopted for human therapeutics will create "bottlenecks" in research and development (Contreras & Sherkow, 2017, p. 1). This raises the possibility of similar problems with access to beneficial medical testing and treatments as occurred with the Myriad case, but on a much larger scale.

7 Conclusion

Is ethical licensing a good thing? In short, no. Patent rights are designed to encourage innovation, rather than to act as ethical checks on research activity. We have explored how ethical licensing is a hindrance to scientific research, it is undemocratic, and it is ineffective as scientists seek to work around its limits rather than take it as an ethical yardstick. In practice, ethical licensing risks reducing engagement with ethics at the research design stage, making ethics a late-stage hurdle to overcome before technology is commercialized.

We note that ethical licensing is only likely to become more complicated and so even less effective. There have been many developments since the foundational CRISPR-Cas 9 technology was patented, and many of these are also subject to patent protection (Scheinerman & Sherkow, 2021, p. 6). As such, there is an increasingly broad base of patent holders who can make decisions about the development of technology (Scheinerman & Sherkow, 2021, p. 6). In this context, there will continue to be ethical questions about what the technology should be used for and who should make these decisions.

The limitations of governance by patents in the CRISPR context indicate that scientists, governments, and the broader public should not take comfort from this approach. The technology should be regulated in a way that upholds democratic legitimacy. Scientists need clear guidance through regulation about what research using CRISPR comes within boundaries deemed acceptable and what should not go ahead. Efforts should be directed to developing appropriate, clear, and consistent regulation, drawing on inclusive processes of public deliberation, and discussions internationally.

Acknowledgments The authors wish to thank the Commonwealth Scientific and Industrial Research Organisation (CSIRO) Advanced Engineering of Biology Future Science Platform for funding this project and the University of Canberra for in-kind support.

References

Australian Government. (2024, April 24). *Gene Drives.* National Gene Technology Scheme. https://www.genetechnology.gov.au/about-gene-technology/gene-drives.

Bell, F. (2016). Empirical research in law. *Griffith Law Review, 25*(2), 262–282. https://doi.org/10.1080/10383441.2016.1236440

Broad Institute. (n.d.-a). *Questions and answers about CRISPR.* https://www.broadinstitute.org/what-broad/areas-focus/project-spotlight/questions-and-answers-about-crispr

Broad Institute. (n.d.-b). *Principles for disseminating scientific innovations.* https://www.broadinstitute.org/principles-disseminating-scientific-innovations

Broad Institute. (n.d.-c). *Information about licensing CRISPR systems, including for clinical use.* https://www.broadinstitute.org/partnerships/office-strategic-alliances-and-partnering/information-about-licensing-crispr-genome-edi

Brownsword, R., & Goodwin, M. (2012). *Law and the technologies of the twenty-first century: Text and materials.* Cambridge University Press.

Contreras, J. L., & Sherkow, J. S. (2017). CRISPR, surrogate licensing, and scientific discovery. *Science, 355*(6326), 698–700. https://doi.org/10.1126/science.aal4222

Cyranoski, D., & Ledford, H. (2018). Genome-edited baby claim provokes international outcry. *Nature, 563*(7733), 607–608. https://doi.org/10.1038/d41586-018-07545-0

de Graeff, N., Dijkman, L. E., Jongsma, K. R., & Bredenoord, A. L. (2018). Fair governance of biotechnology: Patents, private governance, and procedural justice. *The American Journal of Bioethics, 18*(12), 57–59. https://doi.org/10.1080/15265161.2018.1531176

Ezzy, D. (2013). *Qualitative analysis.* Routledge.
Feeney, O. (2019). Editing the gene editing debate: Reassessing the normative discussions on emerging genetic technologies. *NanoEthics, 13*(3), 233–243. https://doi.org/10.1007/s11569-019-00352-5
Feeney, O., Cockbain, J., & Sterckx, S. (2021). Ethics, patents and genome editing: A critical assessment of three options of technology governance. *Frontiers in Political Science, 3.* https://doi.org/10.3389/fpos.2021.731505
Greely, H. T. (2019). CRISPR'd babies: Human germline genome editing in the 'He Jiankui affair'. *Journal of Law and the Biosciences, 6*(1), 111–183. https://doi.org/10.1093/jlb/lsz010
Guerrini, C. J., Curnutte, M. A., Sherkow, J. S., & Scott, C. T. (2017). The rise of the ethical license. *Nature Biotechnology, 35*(1), 22–24. https://doi.org/10.1038/nbt.3756
IP Australia. (n.d.). *Patent Cooperation Treaty.* https://www.ipaustralia.gov.au/international-ip/international-cooperation/the-patent-cooperation-treaty
Ishiguro, K. (2022). *Klara and the sun.* Vintage.
Ledford, H. (2015). CRISPR, the disruptor. *Nature, 522*(7554), 20–24. https://doi.org/10.1038/522020a
McLennan, A. (2018). *Regulation of synthetic biology: BioBricks, biopunks and bioentrepreneurs.* Edward Elgar Publishing. https://doi.org/10.4337/9781785369445
McLennan, A., & Maslen, S. (2023). Impacts of patents on the practice of synthetic biology in Australia: A qualitative study. *Australian Intellectual Property Journal, 34*(2), 1–18. https://search.informit.org/toc/aipj/34/2
National Health and Medical Research Council. (2013). *Australian code for the care and use of animals for scientific purposes* (8th ed.). https://www.nhmrc.gov.au/about-us/publications/australian-code-care-and-use-animals-scientific-purposes#block-views-block-file-attachments-content-block-1
National Human Genome Research Institute. (2025, November 13). *Knockout.* https://www.genome.gov/genetics-glossary/Knockout.
Nicol, D., Paxtonab, R. & Niemeyer, S. with Curato, N., Dryzek, J., Rudge, C., Pemberton, S., & Veri, F. (2022). *Genome Editing: Formulating an Australian community response.* (Occasional Paper No. 12). Centre for Law and Genetics. https://www.utas.edu.au/__data/assets/pdf_file/0011/1634258/OP12-final-report.pdf
Niiler, E. (1999). Terminator technology temporarily terminated. *Nature Biotechnology, 17*(11), 1054. https://doi.org/10.1038/15034
Nordberg, A., Minssen, T., Feeney, O., de Miguel Beriain, I., Galvagni, L., & Wartiovaara, K. (2020). Regulating germline editing in assisted reproductive technology: An EU cross-disciplinary perspective. *Bioethics, 34*(1), 16–32. https://doi.org/10.1111/bioe.12705

Normile, D. (2019). Chinese scientist who produced genetically altered babies sentenced to 3 years in jail. *Science*. https://doi.org/10.1126/science.aba7347

Organizing Committee of the Second International Summit on Human Genome Editing. (2018, November 28). *Statement*. National Academies. https://www.nationalacademies.org/news/2018/11/statement-by-the-organizing-committee-of-the-second-international-summit-on-human-genome-editing.

Parthasarathy, S. (2018). Use the patent system to regulate gene editing. *Nature, 562*(7728), 486–488. https://doi.org/10.1038/d41586-018-07108-3

Patent Cooperation Treaty, June 19, 1970, https://treaties.un.org/doc/Publication/UNTS/Volume%201160/volume-1160-I-18336-English.pdf

Rozen, I. (2016a, September 29). Licensing CRISPR for agriculture: Policy considerations. Broad Institute. https://www.broadinstitute.org/news/licensing-crispr-agriculture-policy-considerations.

Rozen, I. (2016b, December 16). Licensing of IP to maximize public benefit. Broad Institute. https://www.broadinstitute.org/blog/licensing-ip-maximize-public-benefit.

Scheinerman, N., & Sherkow, J. S. (2021). Governance choices of genome editing patents. *Frontiers in Political Science, 3*, Article 745898. https://doi.org/10.3389/fpos.2021.745898

Sherkow, J. S. (2025). A 'bare hope of a result': The second CRISPR patent appeal. *The CRISPR Journal, 8*(5), 317–320. https://doi.org/10.1177/25731599251361 3

Sherkow, J. S., Adashi, E. Y., & Cohen, I. G. (2021). Governing human germline editing through patent law. *JAMA, 326*(12), 1149–1150. https://doi.org/10.1001/jama.2021.13824

The Jackson Laboratory. (2025). *CRISPR/Cas limited use license terms with the Broad Institute*. https://www.jax.org/about-us/legal-information/licenses/crispr-cas-license-terms-broad-institute

CHAPTER 11

The Price of Safety: The Ethics of Occupational Risk Monetization in Brazil

Carolina Bonemer Cury and Caio Caesar Dib

1 Introduction

In risk management, risk monetization is the process of estimating the monetary value of potential losses associated with a particular risk. In the context of Brazilian industrial relations, it refers to policies whereby employers or governments pay workers a premium to compensate for hazardous or unhealthy working conditions (Barros, 2013). The idea of special premiums for unpleasant or dangerous work originates from the economic theory of compensating wage differentials. According to this theory, the labor market, which is assumed to be governed by principles of rationality and balance, provides additional remuneration for occupations involving unfavorable or dangerous conditions (Tavares, 2021). Under free market conditions, it is assumed that well-informed workers would avoid risky jobs unless they were accompanied by a pay differential that

C. B. Cury (✉)
University of São Paulo, São Paulo, Brazil

C. C. Dib
University of São Paulo, São Paulo, Brazil

Sciences Po Law School, Paris, France

J. Hayes, S. Maslen (eds.), *Ethics in the doing of technoscience*,
https://doi.org/10.1007/978-3-032-20002-0_11

compensated for the risk (Shrader-Frechette, 2002). This would compel employers either to make jobs safer or to offer higher wages to attract the necessary number of workers willing to accept such conditions.

Interpreted as a specific form of compensating differential, the monetization of risks would appear as a natural phenomenon of contemporary capitalistic dynamics. However, in the Brazilian legal system, the monetization of risks extends beyond the purely economic sphere. As Nogueira and Gomes (1987) argue, it is a deliberate public policy through which the state simultaneously burdens employers who expose their employees to risks and compensates workers financially for the dangers they face. Thus, there would be two purposes to this policy: first, to financially compensate workers for the risks they take and any damage they suffer and, second, to encourage employers to improve working conditions and avoid activities that generate health risks, so as to avoid further costs (Nogueira & Gomes, 1987).

In other jurisdictions where risk monetization is not in place, occupational safety and health (OSH) decision making is not driven by free market ideals alone but by consideration of the maximum risk to which workers should reasonably be exposed (Hood, Rothstein & Baldwin, 2004). In this case, there is a direct link between employees' conditions of employment and the work of OSH professionals (Aven, 2016). Such professionals conduct risk assessments and make recommendations to management for hazard mitigation in line with legal requirements (as a minimum). As such, OSH professionals have a direct and important role in controlling working conditions and the hazards to which workers are exposed, while in Brazil such professionals are usually sidelined and absent from the negotiating table when OSH decisions are taken. This raises important ethical dilemmas. Examples include the asymmetry of information between employees and employers regarding the risks, and the economic vulnerability resulting from employees' dependence on work for survival. In this sense, it is often argued by Brazilian legal doctrine that, in a context of low wages, health and safety premiums are an important source of income for many low-paid workers. Consequently, from the perspective of workers, beyond merely trying to compensate for risks, this policy creates a conflict between earnings and the preservation of health and safety, including in collective labor relations. At the same time, from the perspective of employers, cheap health premiums may seem like an easier and cheaper short-term solution than the improvement of labor conditions. In this context, the prevention of occupational diseases may become secondary to economic interests and needs from the perspective of upper management, and technical staff such as occupational safety technicians, safety engineers and occupational doctors may be sidelined and left with little power to adequately manage risks. Such cases raise questions about what we mean by ethics in relation to hazardous work, and the status of knowledge and expertise in relation to its management.

This chapter discusses the ethics of upper management professionals of the waste collection sector in Brazil based on a case study of a collective bargaining agreement (CBA) of the sector from the city of São Paulo, Brazil. It will begin with a brief overview of the monetization of occupational risk and collective bargaining in Brazil before introducing our selected case: the CBA of São Paulo's waste collecting workers. We then present the upper management workers as the main moral agents of OSH matters, given the diminished autonomy of OSH professionals, as well as which criteria may be adopted to ethically evaluate their conduct based on communitarian ethics. Afterward, we examine the relationship between community-building and consent to better understand the ethical behavior of the community of waste collection workers. The final sections explore the path for ethical decision making by the upper management regarding the monetization of health and safety risks, emphasizing the need to limit risk monetization and to foment the inclusion of OSH professionals in the community of workers as individuals equipped with essential resources for the improvement of the wellbeing of the labor force.

2 A Snapshot of Occupational Risk Monetization in Brazil

2.1 A Brief Overview of Risk Monetization and Occupational Safety and Health

Brazil's regulations on OSH have many sources and stem directly from the 1988 Federal Constitution. Article 7, XXII of the Constitution determines that the reduction of work-related risks through health and safety rules is the right of all workers. Conversely, risk monetization is also firmly embedded in the legal system, as set out in Article 7, XXIII of the Constitution, which stipulates that premiums must be paid for work carried out in hazardous or unhealthy conditions. Beyond the Constitution, Chapter V of the Consolidation of Labor Laws ("Consolidação das Leis do Trabalho"—CLT) establishes the basis of health and safety protections, which are further detailed in 36 highly technical ordinances known as

"Normas Regulamentadoras" (NR) that set out the rights and responsibilities of employees and workers regarding OSH. NR 15 in particular establishes threshold limit values and other criteria for determining the safety of exposure to chemical, biological, and physical agents. If exposure exceeds the safety threshold as defined by NR 15, the working environment is deemed "insalubre" (unhealthy) and a health premium must be paid, or the risk must be reduced to acceptable levels. Health premium values ranging from 10% to 40% of minimum wage are applied to different hazards as set out in a list established by NR 15.

In practice, legal doctrine acknowledges that employers currently have the option of paying the health premium or improving labor conditions (Daroncho, 2012). Employers may expose workers to risks that are beyond what is deemed safe to their health (the limits established by NR 15) as long as they pay these premiums,[1] which compromises the capacity of even government inspectors to enforce OSH rules, as the adequate payment of a premium means an employer is acting in accordance with the law. Legal doctrine contests the validity of this interpretation, as paying the premium is often more economically advantageous for employers than improving labor conditions, which would hinder the preservation and improvement of OSH (Villatore & Campagnoli, 2015). In any case, the health premium is paid to compensate for the risk, not the damage itself; workers are entitled to further compensation from their employers if they concretely suffer damage due to occupational exposure to harmful agents. This, of course, does not reduce the damage that was done and reflects the manner through which some of our legislation and, especially, our workplace culture regard workers' safety and health as fungible with money, instead of focusing on preventative measures.

2.2 Occupational Safety and Health Enforcement in the Workplace

Compliance with OSH regulations is a duty of all employers, with the day-to-day activities generally carried out by specialized technical staff composed of occupational safety technicians ("Técnicos de Segurança do

[1] There are limits to this. For certain agents, the legislation specifies a "maximum value" above which no worker can be exposed, since this would represent grave and imminent danger to their lives. Furthermore, this applies only to certain health hazards as described by NR 15, not all OSH regulations.

Trabalho"—TST), occupational safety engineers, and occupational doctors and nurses. These technical staff members are mandatorily employed by employers who have a certain number of workers and/or who conduct risky activities, in accordance with rules set out by NR 4.

Although this technical staff holds the expertise to guide employers and employees toward improved safety conditions, in practice, their position as subordinated employees limits their capacity to act and make important risk management decisions (Inoue & Vilela, 2014). Technical staff in Brazil are not entitled to special protections against arbitrary firings and are often not present in collective bargaining or other decision making settings. In fact, it is not uncommon for them to be relegated to formal compliance tasks and to a limited enforcement of rules established by the upper management. Thus, if upper management or a collective bargaining agreement establish that health premiums should be paid and/or no alterations should be made to dangerous working conditions, or even that certain OSH violations will not be corrected, technical staff are often unable to contest this.

For these reasons, actual enforcement of OSH regulations generally falls to the national labor inspection system, whose specialized labor inspectors have the power to inspect, fine, or even halt establishments suspected of violating OSH regulations. However, the effectiveness of these inspections is limited due to the small number of labor inspectors in the country when compared to the number of workplaces to be inspected and the lack of adequate resources, not to mention the fact that these government workers have other duties beyond OSH inspection, such as acting to eradicate child and forced labor (Vasconcelos, 2014). Furthermore, the existence of health premium bonuses limits the capacity of labor inspectors to enforce OSH rules, as we will see below. In this context, upper management decisions and collective bargaining agreements have a far greater impact on the level of employee exposure to dangerous agents than the actions of either technical staff or labor inspectors, which justifies or focuses on the ethics of upper management when dealing with these issues.

2.3 *Industrial Relations and Collective Bargaining*

Brazil's current industrial relations system does not differ greatly from the corporatist model initially implemented in the 1930s, which severely limited freedom of association (Cardoso & Lima, 2020). Although some improvements have been made since then, Brazil has still not ratified the International Labour Organization's (ILO) Fundamental Convention No. 87 on Freedom of Association and Protection of the Right to Organise, 1948, and it does not comply with many of its standards. For instance, despite the current Federal Constitution of 1988 being a social constitution that enshrines a wide range of fundamental rights and freedoms, including freedom of association in trade unions, it still applies the "trade union unicity" rule.[2] According to this rule, only one trade union per professional category can exist in a given geographical area, and this area cannot be smaller than a municipality.[3]

The sole existing trade union is responsible for conducting collective bargaining with employers and their organizations. The negotiated terms then apply to all workers under that trade union's jurisdiction, including those who are not members of the union and who might otherwise disagree with the terms. Until 2017, trade unions could only negotiate terms that were more favorable than those already set out in labor laws. This changed with the 2017 labor law reform, which enabled parties to negotiate reductions in existing legal protections on most subjects.[4] While OSH rules cannot be reduced through collective bargaining, the number of working hours and the frequency and length of breaks can be negotiated. It is also now possible for a trade union to authorize overtime in activities that present health risk, when, before, only the competent authority could grant this permission after conducting an inspection to confirm overtime could be done safely. Even more importantly, although it is not possible to negotiate the non-payment of occupational health and hazard premiums when they are due, it became possible to negotiate the value of health risk premiums, which previously could only be established through technical appraisal of the risks involved.

[2] Unicidade sindical, in the original Portuguese.

[3] The Committee on Freedom of Association of the ILO has stated that the "trade union unicity" rule "violates the right of workers and employers to establish and join organisations of their own choosing" in two separate cases against Brazil, No. 2523 of June 2007 and No. 2739 of 2010 (International Labor Office, 2018).

[4] According to article no. 611-A of the CLT.

3 Case Study: Collection of Non-hazardous Waste in Brazil

The case we address in this chapter includes several elements that are relevant for our analysis on the ethics of risk monetization: (1) the greater vulnerability of workers employed in this activity to occupational accidents and diseases; (2) the fact that waste collection is an activity that suffers from low social prestige and predominantly employs vulnerable individuals from low-income households who had limited access to education; and (3) the nature of the collective bargaining clauses that were accepted by trade unions in negotiation with employer organizations from this sector.

In Brazil, workers employed in the collection of non-hazardous waste consistently top the charts of occupational accidents and diseases. In 2024, these workers represented 1.05% of all Social Security leaves motivated by an occupational accident or disease. Moreover, they are listed as a priority group for the improvement of OSH conditions by a key government-led project (SmartLab, n.d.). This data is unsurprising once we realize the way waste collection is conducted in Brazil and the working conditions to which these workers are subjected. Although technological development has allowed the automation of several stages of waste collection in many countries, waste collection in Brazil remains a form of hazardous manual labor. Workers generally ride on narrow platforms (called *estribos*) on the back of garbage trucks around cities, climbing up and down to grab garbage bags directly and throw them into the waste compressor of the truck. For this reason, workers in this sector are likely to suffer cuts and punctures from perforating-cutting objects inside garbage bags, contusions and crushing from moving parts of the truck, and sprains and falls from running and climbing up and down the truck platforms (Robazzi et al., 1994). They are also constantly exposed to chemical and biological agents present in garbage that present health risks. As for socio-economic vulnerability, research conducted by Belarmino and colleagues indicate that over 60% of the waste collection workers interviewed had not finished high school; it also shows that these workers are aware of the social stigma and prejudice linked to their occupation, even though they also recognize that their jobs are socially relevant (Belarmino et al., 2022).

To investigate these issues, we conducted a document analysis of a single CBA to examine the ethical and institutional dynamics of OSH negotiations. The CBA selected for analysis is a 2019 agreement signed by

SIEMACO (the trade union representing waste collection workers of the city of São Paulo) and SELUR (the employers' organization of the waste collection sector).[5] This particular CBA was chosen due to the fact that it was signed in after the 2017 labor law reform and includes clauses in which health premium values were negotiated, but before the government established specific safety regulations for the waste collecting sector (NR 38), in 2022. Thus, although it is no longer in force, it gives us a snapshot of OSH negotiations after the flexibilization of collective bargaining rules and allows us to observe the outcome of these negotiations. Our analysis focuses on: (1) its temporal positioning—negotiated between the 2017 labor law reform and the 2022 adoption of sector-specific OSH regulations—and (2) its content, which lacks the technical preventive measures that would be necessary in a high-risk sector like waste collection.

4 Findings

4.1 The Collective Bargaining Agreement

Most, if not all, of this CBA's clauses pertaining to OSH held a compensatory nature, rather than a preventive one (Cardoso & Lima, 2020), giving an indication of how deeply entrenched the risk monetization policy is. However, there are some clauses that are particularly pertinent, which we list in Table 11.1.

It is also crucial to highlight that we consider the clauses listed above to be concessions from the trade union to the employers' organization, instead of an advancement of labor rights. The main reason for this is that health premiums need not be negotiated to be guaranteed; indeed, if the employer does not reduce worker exposure to unhealthy agents or pay the correct premium amount, they can be fined by the labor inspection agents or sued by individual workers or by the trade union. Moreover, according to Brazilian occupational health regulations, activities involving permanent contact with garbage, in both collecting and processing stages, should be compensated by a 40% premium. Therefore, the negotiation of 10% and 20% premiums for certain workers, with consequent withdrawal of a lawsuit, is likely to be a concession from the trade union.

[5] This CBA may be accessed by searching the registry number "SP000257/2020" in the official Ministry of Labor database of CBAs: http://www3.mte.gov.br/sistemas/mediador/

Table 11.1 CBA clauses for collection of non-hazardous waste. Data from the CBA between SELUR and SIEMACO

Clause numbers	*Subject*
11	This clause sets the value of health risk compensation that must be paid to different types of employees. It establishes 10% premiums for helpers, 20% premiums for cleaners and workers employed in waste treatment plants and landfills, and 40% for waste collectors and utility hole workers. It establishes that the trade union will withdraw judicial claims against employers demanding health premiums for helpers. It establishes that the payment of a health premium to workers does not represent a recognition, by the employers, of the existence of unhealthy working conditions. It establishes that the payment of a health premium shall not be an impediment to the realization of overtime. It establishes that the payment of a health premium will not automatically result in special retirement rights for employees.
35	The trade union authorizes overtime on activities that present risk to the health.

The exposure to unhealthy working conditions might also result in special retirement benefits for employees and in additional obligations on the part of the employer, particularly regarding periodical medical exams and extra social security charges and can impact the burden of proof in individual or collective lawsuits pertaining to occupational accidents and diseases. Thus, the clauses establishing that employers do not recognize the unhealthiness of the working environment despite paying a health premium represent yet another concession on the part of trade unions and can reduce employers' future obligations (and employees' rights) regarding OSH.

The lack of technical precision in this CBA becomes evident when we compare it to NR 38, a specific sectoral OSH regulation that was approved in 2022. NR 38 contains specific clauses pertaining, among other things, to ergonomic risks, exposure to noxious gas from the garbage truck, and speed limits for trucks whenever workers are riding on the outside. These important OSH issues were well known to OSH professionals and could have been regulated before 2022 by CBAs, but were left untouched. The absence of specialized technical staff at the negotiating table may help explain this omission. While the failure to address these risks could indicate that either employers and trade unions could not reach an agreement on them or they did not wish to regulate them, it could equally indicate that social partners did not have the technical skills to adequately identify, measure, and regulate all relevant risks. Elaborating OSH clauses is a difficult task that requires specialized knowledge so that new rules do not have unintended consequences from economic, legal, and safety perspectives. For instance, establishing a speed limit for trucks could result in more trucks being needed to cover the same areas in the same time span, and extra safety equipment could actually introduce new risks, if not properly analyzed in the context of the activity. However, the lack of independence of technical staff and their non-participation in these contexts leaves negotiation and decisions mostly to upper management.

4.2 *Upper Management as the Center of Ethical Decision Making*

As we highlighted in the previous section, while technical staff are assigned the task of enforcing OSH regulations in the workplace, in the Brazilian context they often lack proper autonomy to fulfill their intended role. They share the status of other workers and many of the experiences of other employees. They are often excluded from any important decisions which may concern a company, such as collective bargaining and OSH policies. In the absence of empowered technical staff, upper management and unions represent the parties directly involved in CBA negotiations and so should be considered relevant moral agents, which leads to the need for them to fill the role regardless of the lack of technical expertise. Upper management professionals are first and foremost agents of their company, who in turn are counterpoised by unions as representatives of the collectivity of employees. These identities—as well as their ethics—are born dialogically, as a result of a process of recognition of their respective social roles (Taylor, 1991).

Labor relations are known for their innate asymmetries between employer and employee. Economic power, information, and institutional support are notably unequal between the two parties. Legal and institutional measures in labor relations take shape to enhance economic liberty: as Hale (1943) points out, these are necessary to ensure a certain degree of freedom of contract by the weaker parties. Thus, legal and institutional restraints operate as a counterbalance to economic power. It makes sense for workers to seek unionization grounded in economic aspects of labor.

It has been shown to improve wages and amenities in the long run (Lagos, 2024), countering circumstances of monopsony and other asymmetries which impact bargaining power, such as information asymmetry in capital markets (Xing & Yan, 2018). Information asymmetry limits freedom of choice (Akerlof, 1970) and may turn risk monetization from optional into mandatory. Labor unions are often thought to act on power and struggle, aiming to meet at less unequal conditions of negotiation the bargaining power held by employers and thus achieve better salaries.

Labor unions legitimize their existence to workers by defending their interests in the shape of better salaries and work conditions. Unions also may be justified normatively due to the innate incompleteness of employment contracts (Lindblom, 2019): information asymmetry and coercion mean the consent we take as basis for contracts is always under duress, and unionization provides contestability to a collective of workers. Unions give voice to workers and make the playing field less uneven (Lindblom, 2019, p. 201). Hence, they allow workers to respond to the disadvantageous scenario they initially see themselves in. Their role is to be a mediating object which reduces coercion and improves communication and the capacity to obtain and understand the conditions of the employment relationship.

However, the role of unions has been systematically undermined in the last decade. Brazilian trade unions have had steadily declining membership rates for years (ILOSTAT, 2024). Aside from this, it indicates potential representativity and legitimacy issues (Reis, 2021). This has recently turned into a financial issue for trade unions. Until 2017, membership rates did not affect a union's income, due to the so-called trade union tax—a mandatory contribution that was automatically deducted from every worker's wage and directed to their union, in spite of their membership status. The 2017 labor law reform revoked the mandatory nature of this contribution, allowing workers without union membership to oppose wage deductions. According to Batista and Seferian, trade unions' incomes dropped by around 90% shortly after this reform (Batista & Seferian, 2020), partially due to a concerted effort by employers to convince or force employees to oppose these payments through anti-union practices (Ministério Público do Trabalho, 2025). This brutal cut in financial resources was accompanied by the new possibility of rights being reduced through collective bargaining, also brought in by the labor reform. This forced the weakened unions to use whatever resources they had to avoid rollbacks, which was not always successful.

The waste collecting sector's CBA studied here was the first one negotiated after these changes were implemented by the legislation. Although we cannot affirm that certain clauses were negotiated only because the trade union had been weakened, it is possible to identify that, in clause 11, the trade union agreed to give up ongoing judicial causes and to pledge not to file other lawsuits on the theme of unhealthy work premiums. This may indicate that a previously belligerent trade union was forced to change its strategies and agree to less advantageous terms due to the changed legal landscape.

As a result of the 2017 labor reform reinforced an atomist and instrumentalist stance in labor relations in Brazil (Taylor, 1991, p. 111). Hence, power has shifted in favor of the employer, which has become the force which dictates CBAs and one-sidedly shapes policies for the work environment. That means the decision making instances of companies, i.e., senior white-collar workers and upper management professionals, are currently the main agents where occupational risk is concerned. Safety and hazard pay is one of many decisions within their competence, in contrast with the praxis which tend to responsibilize blue-collar workers for their own safety (Oliveira, 2007).

Given the protagonism of upper management in dictating OSH norms, we should outline how to evaluate the ethics of those white-collar professionals in this case. Companies are human-made objects. When speaking of these objects, Bruno Latour (2005, p. 74) states that "[i]f they were as weak as the social skills they have to reinforce (...) what gain would they be?" Social ties based on individuals are often weak and volatile. With the goal of making them durable and reliable, collectives fabricate institutions, books of rules and procedures. In this sense, they are an institution in its truest sense, also behaving as decision makers and providing identity while seeking their respective objective (generating wealth, mediating relationships between stakeholders, filling a social role) and being a cognitive apparatus which deeply influences the choices and the outlook of its members.

Companies tend to emphasize the instrumental reason and disengaged demeanor (Taylor, 1991, pp. 103–104) as guidance for their behavior, such as the reduction of transaction costs (Coase, 1937). With this comes the attempt to isolate the company (Coase, 1937) from its externalities and social costs—first through its economization and then through the formatting of the institution and the market as such (Callon & Latour, 1997, pp. 7-8). But it is worth recalling that identity and the ethics framework are inherently dialogical and dependent on the social roles of actors. Companies were fabricated and nurtured as legal institutions due to their ability to provide economic governance which services the community, making managers trustees to society as a whole (Berle, 1932). In this sense, managers are actants supposed to align with the collective values and interests of stakeholders. If these actors are thought to have a certain degree of agency, we may expect them to champion the substantive morality (Fuller, 1969) of the collective they represent. This can be done through CBAs, which crystallize working standards the various actors deem fair, safe, and reasonable.

As we argued, the weakening of unions led to the strengthening of employers and their agency. Employers now have greater bargaining power to force unions to accept disadvantageous collective rules, including in some aspects that affect OSH. A case can be made for the extension of their moral responsibility as well, given how they now negotiate with a particularly frail party. Now companies themselves become ontologically connected to the protection of the interests of employees and in this sense have an ontological responsibility (Jonas, 1984) toward them.

The CBA we presented above seems to indicate that companies and labor unions apparently agree on the monetization of health and safety. Nonetheless, we should assess the CBA with suspicion, given the presence of instrumental reason in their conception and the weakened status of unions as organizations oriented toward the protection of the interests of workers.

4.3 *Protecting the Interests of Workers*

Having established that unions and upper management are the moral actors in negotiation of the CBA, it is relevant to consider what workers want from the CBA and the extent to which unions may represent workers' interests. Workers in waste collection tend to come from low-income households meaning that the financial compensation received is a strong factor in the value they place on their job (Kulkarni, 2000, p. 220). In addition, their typically low education level naturally restrains freedom of choice and control over one's labor (Braverman, 1998 [1974]), making them more vulnerable to monopsony exploitation (Boal & Ransom, 1997). The vulnerability of waste collection workers often leads to mirroring the transactional behavior of employers, adopting the ethical framework of businesses by perceiving their resources (i.e., organized workforce) as a means for leverage in negotiations for better payments rather than a goal in itself, whose wellbeing should be prioritized (Briefs, 1952, pp. 85–86).

However, perceiving bargaining practices solely on monetary terms may obfuscate the much broader range of demands of these groups for better working conditions, such as human dignity, workplace safety standards, and a sense of belonging (Hordijk, 1978, p. 46). What we see, then, is an apparent contradiction between two different dispositions toward bargaining in labor relations: as purely transactional and competitive, centered on maximizing income and economic advantages, and as existential (but not necessarily collaborative), a way of improving work environments and changing fundamental aspects of labor-management relationships.

This confusion of priorities falls in the scope of the ever-present information asymmetry in labor relations, which may be more relevant than usual when the activity involves particularly impactful consequences to a community's environment (Kulkarni, 2000, p. 218). In the present case we see two different instances in which asymmetry occurs: (1) between the firm and the collective of employees, in a more traditional sense (Akerlof, 1970), as well as different in the capacity to process the information available—that is, "cognitive asymmetry" (Gouvêa, 2022); and (2) within the firm, the union, and the collective of workers themselves, as a form of internal asymmetry.

These two forms of asymmetry intertwine, enabling opportunistic behavior. They influence decision making by affecting the competence of agents and reinforce atomization and isolation within communities, weakening political action (Taylor, 1991, p. 117). Decision making is affected by this asymmetry mostly in the form of embracing riskier behavior, based on the belief that the parties will not be affected by potential negative effects of their choices (Arrow, 1963) through adverse selection and moral hazard. In this case, we may observe adverse selection occurring when workers, for any reason, lack sufficient support from technical staff and have limited or wrong information concerning risks innate to waste collection. Thus, they accept greater flexibility of health and safety rules.

On the other hand, moral hazard appears in how the CBA shows a substantive reduction in health and safety premiums, as well as the favoring of compensatory measures in place of preventive ones. In fact, the CBA does not contain any relevant preventive measures, aside from a clause requiring employers to provide sunscreen, even though waste management is considered to be a high-risk activity with high rates of occupational accidents and diseases.

Internal asymmetry may also affect contestability within the community of waste collection workers, what impacts how divergent voices are heard both internally and externally (Lindblom, 2019, p. 200). This form of asymmetry works as a disintegrating force, further weakening the sense of belonging within workers employed in different roles with diverse perceptions of risk and undermining the inclusion of OSH technicians beyond the formal governance structures, in which they already suffer from isolation. This becomes autopoietic at a certain point, with isolation driving further both kinds of asymmetry and further jeopardizing the social ties.

Community-building and information are intrinsically related, playing an important role in preventing opportunistic behavior and fragmentation. Meanwhile, escaping discrimination (particularly against OSH professionals) is necessary to reduce these asymmetries. Hence, upper management has a key role in properly inserting OSH actors in policies and the corporate environment. Considering the legal obstacles to imposing their presence in collective bargaining, it becomes even more necessary to avoid their isolation and empower their actions in order to make their agency something desired by the community of workers.

5 Ethically Oriented Health and Safety Standards

Throughout this chapter we discussed the ethical dimension of decisions and its potential limitations in collective bargaining in non-hazardous waste management, framing it from the perspective of upper management. It should be noted that the exercise of consent and contestability by communities of workers faces many challenges: some derive from the weakening of unions particularly since the deregulation of Brazilian labor law in 2017; others are a result of environmental factors and constituent features of its members. While it is important to recognize that the agency of upper management is limited by the external constraints of legal and social nature, we also pointed out that white-collar workers still have the most agency as actors who hold the power to shape corporate policies and influence worker unity.

We have seen that much of the issue lies in (1) information and cognitive asymmetries and (2) the use of bargaining power to aim for higher salaries. When it comes to this first element, there is much to be gained by empowering OSH professionals policy-wise and community-wise. OSH technicians must be put in a position where they take active participation in developing policies which pertain to the labor environment, restoring their authority to act upon their competence. Likewise, they should hold a pedagogical role toward other employees, raising awareness of the importance of upholding health and safety beyond immediate benefits.

However, the second elements indicate that sticking to the CBA and to standard regulations may not be enough. The social vulnerability of waste collection workers in Brazil leads a race to the bottom concerning risk monetization, in which salary increase is sought after whenever possible. Corporate policies should operate in a *soft ethics* manner, seeking complementarity and improvement of regulatory standards (Floridi, 2021, p. 621)—state regulations and OSH norms, the CBA, and labor law provisions. After all, health and safety are supposed to be institutional goals orienting the conduct of managerial positions.

In contrast, we see that the practice of hazard control in our case study has been affected by the priorities of workers who come mostly from low-income households and to whom monetary compensation may represent a very significant improvement in living conditions. For this reason, their health and safety are far from being the workers' highest priorities. This is reflected in the position taken by unions in negotiating the CBA on the workers' behalf and reinforced by the union's lack of access to competence resources when it comes to workplace health and safety issues. It has also been impacted by the asymmetries which are present between waste management firms and employees.

In general, this situation also emphasizes how sensitive a community of workers can be to upper management conduct where choices by white-collar professionals may either deepen isolation and politicization, weakening further the sense of belonging among employees, or endorse common action and restore the authenticity of the workforce. Either way, it has the potential to lead to a vicious or virtuous cycle (Taylor, 1991, p. 118).

The case study illustrates this may reflect in CBAs as the propensity to opt for less stringent rules when it comes to safeguarding the integrity of workers in favor of monetary compensation. Disengaged, instrumental reason impacts how voices are heard, given how information asymmetries may happen within companies, which in turn affects the capacity of workers to contest the incompleteness of labor contracts, and may even support indifference toward explicit violations of labor law (Friedman, 2016, p. 183) as long as they satisfy short-term interests. Hence, salary increases are chosen over OSH standards whose negotiability is questionable at best and which is deeply affected by information and cognitive asymmetries.

Labor contracts represent a continuum rather than an event (Llewelyn, 1931, p. 747), with the union acting as a prophylactic regulator and a mediator providing completion to the contract. However, information asymmetry reinforces that unions are often handicapped when negotiating the livelihood and safety of their workers. This means that decisions such as those made during collective bargaining have long-term implications that are disregarded. And in an environment where social conventions tend to be even more relevant than the legal order, the weight of decisions increases. The unforeseen consequences have the potential to impact beyond a single labor contract and reach the collectivity of workers for a prolonged time.

The principle of precaution should be exercised due to the acknowledgment of health risks of the working conditions of waste collection. Jointly to the burden of proof (Science and Environmental Health Network Staff, 1998), we may speak of a duty regarding posterity (Jonas, 1984, pp. 38–41) of upper management toward their trustees—that is, society in general, including the workers under their supervision. This of course includes the empowering of OSH technicians and their inclusion in the spaces of deliberation in order to overcome current notions of risk monetization. After all, it can affect both the immediate wellbeing of employees and society in general, as it propagates the normalization of putting a price on safety as a manner of salary increase.

6 Final Remarks

The monetization of occupational risk in Brazil highlights the conflict between socio-economic vulnerabilities and the assertion of the fundamental rights to health and safety at work. Hazard pay has been often used as a supplementary source of income while seeming as a practical way to balance dangerous working conditions, but in practice it often masks a structural failure to protect the best interests of workers. Prima facie agreements of risk monetization are particularly problematic when decisions are taken in a disengaged manner and without the guidance of technical personnel, with upper management and unions opting for putting a price in safety while a fragmented community of workers, especially those in low-paid, high-risk occupations, accept hazardous conditions for an increase in income.

The fragile position of labor unions and the lack of agency of OSH staff mean that upper management holds the upper hand in decisions concerning risk monetization. It puts these white-collar professionals in a challenging role, given that their status as trustees of multiple stakeholders clashes with the dominant instrumental reason which favors the monetization in question. The case studied exemplifies how workers can find themselves in a precarious position between advocating for both fair compensation and safer working environments and internal and external shortcomings. This communitarian analysis reveals that, while acting as a moral agent, upper management has an essential task in finding solutions amidst the significant limitations in the ability of other actors to make ethical decisions under conditions of market asymmetry, legal ambiguity, and persistent economic vulnerability.

These are structural challenges that can only be dealt with by facing the de facto absence of technical guidance behind decisions. By giving voice to OSH personnel and striving for their inclusion within the community of workers, those in managerial positions may improve decision making by ensuring agents are better equipped for understanding the implications of their choices and how to better align them to health, safety and precaution goals.

Employer accountability—both toward and beyond regulatory standards—and the empowering of OHS professionals equipped to ensure the wellbeing of employees may be the first steps in guiding the quest for ethical standards of health and safety at work beyond putting a price on safety.

References

Akerlof, G. A. (1970). The market for “lemons”: Quality uncertainty and the market mechanism. *The Quarterly Journal of Economics, 84*(3), 488–500. https://doi.org/10.1016/B978-0-12-214850-7.50022-X

Arrow, K. J. (1963). Uncertainty and the welfare economics of medical care. *The American Economic Review, 53*(5), 941–973. https://www.madewill.com/wp-content/uploads/2020/10/Arrow632.3.pdf

Aven, T. (2016). Risk assessment and risk management: Review of recent advances on their foundation. *European Journal of Operational Research, 253*(1), 1–13. https://doi.org/10.1016/j.ejor.2015.12.023

Batista, F., & Seferian, G. (2020). Financiamento sindical, direito do trabalho e crise: Aproximações empíricas ao tema da funcionalidade do direito do trabalho para o capitalismo. *Revista Direito e Práxis, 11*(4), 2669–2695.

Belarmino, D. V. B., Pagani, M. E. B., Tanouye, A. T. A., Garcia, L. C., & Massuda, E. M. (2022). Trabalho e saúde: Percepção de coletores de lixo. *Revista Brasileira de Medicina do Trabalho, 20*(4), 574–581. https://www.rbmt.org.br/details/1740/pt-BR/trabalho-e-saude%2D%2Dpercepcao-de-coletores-de-lixo

Berle, A. A., Jr. (1932). For whom corporate managers are trustees: A note. *Harvard Law Review, 45*(8), 1365–1372. https://www.ius.uzh.ch/dam/jcr:3a26204d-17fa-401e-9e3c-dea23049159a/Berle_For_whom_are_corporate_managers_trustees_note.pdf

Boal, W. M., & Ransom, M. R. (1997). Monopsony in the labor market. *Journal of Economic Literature, 35*(1), 86–112.

Braverman, H. (1998). *Labor and monopoly capital: The degradation of work in the twentieth century* (25th anniversary ed.). Monthly Review Press.. (Original work published 1974).

Briefs, G. A. (1952). The ethics of labour-management relations. *Industrial Relations, 7*(1–2), 81–106.

Callon, M., & Latour, B. (1997). "Tu ne calculeras pas!" ou comment symétriser le don et le capital. *La revue du MAUSS, 9*, 45–69.

Cardoso, A. C. M., & Lima, C. R. (2020). A negociação coletiva e as possibilidades de intervenção nas situações de risco à saúde no trabalho. *Revista Brasileira de Saúde Ocupacional, 45.* https://www.scielo.br/j/rbso/a/MsY5LTzXJmqJ7TGmZ7xTNsq/abstract/?lang=pt

Coase, R. H. (1937). The nature of the firm. *Economica, 4*, 386–405. https://doi.org/10.1111/j.1468-0335.1937.tb00002.x

Daroncho, L. (2012). Saúde laboral: O adicional de insalubridade e o direito fundamental ao meio ambiente de trabalho hígido. *Caderno Ibero-Americano de Direito Sanitário, 1*(1), 44–66.

Barros, A. M. d. (2013). *Curso de direito do trabalho* (9th ed.). LTr.

Floridi, L. (2021). The end of an era: From self-regulation to hard law for the digital industry. *Philosophy & Technology, 34*(4), 619–622. https://doi.org/10.1007/s13347-021-00464-w

Friedman, L. (2016). *Impact: How law affects behavior*. Harvard University Press.

Fuller, L. L. (1969). *The morality of law* (Rev. ed.). Yale University Press.

Gouvêa, C. P. (2022). *A estrutura da governança corporativa*. Quartier Latin.

Hale, R. L. (1943). Bargaining, duress, and economic liberty. *Columbia Law Review, 43*(5), 603–628. https://doi.org/10.2307/1117481

Hood, C., Rothstein, H., & Baldwin, R. (2004). *The government of risk: Understanding risk regulation regimes.* Oxford University Press.

Hordijk, A. (1978). Trade unionism and ethics. In C. van Dam & L. Staalaert (Eds.), *Trends in business ethics: Implications for decision-making* (pp. 43–50). Kluwer.

ILOSTAT. (2024). *Statistics on union membership.* https://ilostat.ilo.org/topics/union-membership/

Inoue, K. S. Y., & Vilela, R. A. G. (2014). O poder de agir dos técnicos de segurança do trabalho: Conflitos e limitações. *Revista Brasileira de Saúde Ocupacional, 39*(130), 136–149. https://doi.org/10.1590/0303-7657000095813

International Labor Office. (2018). Freedom of association: Compilation of decisions of the Committee on Freedom of Association (6th ed.). International Labour Organization. https://www.ilo.org/wcmsp5/groups/public/%2D%2D-ed_norm/%2D%2D-normes/documents/publication/wcms_632659.pdf

Jonas, H. (1984). *The imperative of responsibility: In search of an ethics for the technological age.* University of Chicago Press.

Kulkarni, S. P. (2000). Environmental ethics and information asymmetry among organizational stakeholders. *Journal of Business Ethics, 27*(3), 215–228. https://doi.org/10.1023/A:1006468200761

Lagos, L. (2024). Union bargaining power and the amenity-wage tradeoff (IZA Discussion Paper No. 17034). Institute of Labor Economics (IZA). https://docs.iza.org/dp17034.pdf

Latour, B. (2005). *Reassembling the social: An introduction to actor-network-theory.* Oxford University Press.

Lindblom, L. (2019). Consent, contestability, and unions. *Business Ethics Quarterly, 29*(2), 189–211. https://doi.org/10.1017/beq.2018.17

Llewelyn, K. (1931). What price contract?—An essay in perspective. *Yale Law Journal, 40*, 704–751.

Ministério Público do Trabalho. (2025). MPT investiga atos antissindicais nas regiões de Campinas. *Jundiaí e Piracicaba.* https://www.prt15.mpt.mp.br/informe-se/servidores/2-uncategorised/1803-mpt-investiga-atos-antissindicais-nas-regioes-de-campinas-jundiai-e-piracicaba

Nogueira, D. P., & Gomes, J. R. (1987). Insalubridade: Ainda um desafio. *Revista da Universidade de São Paulo, 5*, 99–106.

Oliveira, F. de. (2007). A persistência da noção de ato inseguro e a construção da culpa: Os discursos sobre os acidentes de trabalho em uma indústria metalúrgica. *Revista Brasileira de Saúde Ocupacional, 32*(115). https://www.scielo.br/j/rbso/a/pNzfHH5wYRztV6KTf43zkFL/?lang=pt

Reis, I. M. (2021). Legitimidade social x legitimidade jurídica: Uma análise sob a perspectiva da crise de representatividade dos sindicatos. *Revista do Tribunal Regional do Trabalho da 3ª Região, 67*(104), 217–237.

Robazzi, M. L. C. C., & Gir, E., & Moriya, T. M., & Pessuto, J. (1994). O serviço dos coletores de lixo: Riscos ocupacionais versus agravos à saúde. *Revista da Escola de Enfermagem da USP, 28*(2), 177–190. https://www.scielo.br/j/reeusp/a/wp5QHK93m5rPdJMD8QKJqPr/?format=pdf

Science and Environmental Health Network Staff. (1998, January 26). *Wingspread conference on the precautionary principle.* https://web.archive.org/web/20051102060553/http://www.sehn.org/wing.html

Shrader-Frechette, K. (2002). Trading jobs for health: Ionizing radiation, occupational ethics, and the welfare argument. *Science and Engineering Ethics, 8*(2), 139–154.

SmartLab. (n.d.). Segurança e saúde no trabalho.. Retrieved May 4, 2025, from https://smartlabbr.org/sst/localidade/0?dimensao=perfilCasosAfastamentos

Tavares, S. G. C. (2021). *Monetização dos riscos no meio ambiente do trabalho: Uma leitura a partir do liberalismo igualitário.* Editora Fi. https://doi.org/10.22350/9786559170951.

Taylor, C. (1991). *The ethics of authenticity.* Harvard University Press.

Vasconcelos, F. D. (2014). Atuação do Ministério do Trabalho na fiscalização das condições de segurança e saúde dos trabalhadores, Brasil, 1996–2012. *Revista Brasileira de Saúde Ocupacional, 39*(129), 86–100.

Villatore, M. A. C., & Campagnoli, A. de F. P. F. (2015). Risco ocupacional e custos: Análise crítica da política de monetização do risco com abordagem de normas do direito espanhol. In *III Encontro de Internacionalização do CONPEDI, Madrid* (Vol. 6).

Xing, X., & Yan, S. (2018). Labor unions and information asymmetry among investors. *Quarterly Review of Economics and Finance, 69*, 174–187. https://doi.org/10.1016/j.qref.2018.02.004

CHAPTER 12

AI as an Epistemic Technology: Navigating Ethics in Practice in Cases of Distributed Decision Making

Sarah Maslen and Jan Hayes

1 Introduction

In recent years, the single biggest change in how professionals work pertains to advances in artificial intelligence (AI) technologies. In this final chapter, we examine the ethical challenges that professionals and industries face as AI technologies come to form part of their decision making processes. There is an established tradition of investigating the ways in which human knowing is entangled with the digital with analyses especially well developed in relation to health practices (Algera, 2023; Fotopoulou & O'Riordan, 2017; Lupton & Maslen, 2018; Maslen & Harris, 2021; Oudshoorn, 2008; Pink et al., 2017). While technoscientific workers are all, in their own ways, entangled with the nonhuman whether we are talking about physicians with stethoscopes (Harris, 2016; Sterne, 2001), scientists and engineers with models (Maslen & Hayes, 2022;

S. Maslen (✉) • J. Hayes
School of Property, Construction and Project Management, RMIT University, Melbourne, VIC, Australia
e-mail: sarah.maslen@rmit.edu.au; jan.hayes2@rmit.edu.au

J. Hayes, S. Maslen (eds.), *Ethics in the doing of technoscience*,
https://doi.org/10.1007/978-3-032-20002-0_12

Myers, 2008), or synthetic biologists with enzymes like Cas-9 (see Chap. 10), AI technologies raise questions in unprecedented ways about how decisions are being made and the degree to which they warrant epistemic trust. Where technologies like the stethoscope are designed and used to aid in the acquisition of knowledge, AI technologies are also designed to deal with epistemic content (i.e., data) and to perform epistemic operations (i.e., analysis, predictions) (Alvarado, 2023).

AI technologies are not singular but encompass a range of computerization advances. Our focus in this chapter is AI technologies characterized by provision of information for purposes and tasks that were "previously reserved for human experts and epistemic authorities" (Hauswald, 2025, p. 56), in contrast to examples that solve practical problems as in the case of self-driving cars. In particular, we are concerned with what is termed "narrow" AI (Dafoe, 2018, p. 19) typically based on Deep Learning approaches (Janiesch et al., 2021). Deep Learning refers to systems that adopt multilayered neural networks to enhance learning abilities more akin to the human brain, in which a system can learn progressively by examining examples in relation to a hierarchy of concepts, rather than requiring task-specific programming (Goodfellow et al., 2016). The applications of Deep Learning are vast, finding a place in various critical contexts including medical diagnostic support (Asif et al., 2025), medical imaging (Islam et al., 2024), cockpit and air traffic control support (Kirwan, 2025), chemical safety (Jiao et al., 2020), and engineering to manage what is acknowledged to be an overly complicated paper trail (Elyan et al., 2020), among others. These contexts in which professionals make high stakes decisions are our particular interest.

It has become a refrain within industry, government, civil society, and in the academy that ethics must be an important consideration in the development and adoption of AI technologies (Coeckelbergh, 2020). We can appreciate the primacy of ethics in the context in which scholarly and public debates have been equally characterized by an optimistic/pessimistic dichotomy. In short, "Either the machines will save us or they will destroy us" (Dahlin, 2024). Much of the work on AI ethics to date has centered on establishing sets of principles, such as Floridi and Cowls' (2022) Unified Five Principles for AI in Society (beneficence, non-maleficence, autonomy, justice, and explicability). There have also been growing calls for a move toward conceptualizing AI ethics in practice, as the principles that have been developed are both theoretical and offer high-level guidelines only (Green, 2021; Kwon & Hong, 2025). While useful to articulate, high-level principles are not grounded in the practical challenges that professionals face, and also risk being treated as an exercise in ethics washing in comparison with a deeper engagement. Moving toward applications of ethics in practice also recognizes that AI "is situated differently in different situations" (Dahlin, 2024).

Epistemic matters are a critical part of professional ethics in technoscientific domains. Professionals are holders of specialized forms of knowledge spanning both the conceptual and embodied (Maslen, 2025; Prentice, 2013; Vertesi, 2015). Practicing competently is a cornerstone of professional codes of ethics, meaning developing and maintaining the domain-specific knowledge and skills to undertake a professional role, acting on the basis of adequate knowledge which can involve seeking the input of others, representing knowledge objectively, and undertaking activities that expand knowledge for the profession such as involvement in training and research (e.g., see Australian Medical Association, 2016; Engineers Australia, 2022). The underlying assumption in claims that AI technologies will make work more efficient is that "this will happen because AI itself will take over some of our decision-making tasks and deliver solutions to problems directly and autonomously" (Buccella, 2025). However, especially in cases of high stakes decision making, technical professionals are often called on to reconcile "competing values, contextual ambiguities, and unforeseeable consequences—dimensions that current AI systems cannot fully grasp" (Shin, 2025). This notion that AI technologies operate autonomously glosses over the continued and often significant work that professionals must do (Carboni et al., 2023). It also raises critical issues of epistemic trust in light of the widely known reality of so-called hallucinations within AI models (Huang et al., 2025).

The following explores these issues of knowing and epistemic trust with a focus on three problems in particular, including those of distance, rare events, and over-reliance. We also address the distinction between non-safety-critical and safety-critical systems, which some sectors have drawn in reasoning through deployment of AI technologies. We conclude with a discussion of whistleblowing, as a last ethical line of defense for professionals, which is perhaps even more important in the case of AI due to the lack of clarity among outsiders over how models operate but is largely unexamined in the existing AI literature (Olesen, 2025; Wu, 2024). We argue that we need to be thinking of AI ethics in practice not only as operationalized principles as others have called for (Kwon & Hong, 2025), but as grounded in the decision making moment. If AI is to support the best decision making, we have to stick closely to this decision making moment and the actors it involves. The goal must be to see not only what AI can or cannot do, but what professionals do currently and must continue to do, and how AI may or may not help with that.

2 Tackling the Problem of Distance

One of the epistemic issues grappled with in this volume relates to what is often termed the "many hands" problem (Thompson, 1980; van de Poel et al., 2015), in which the sheer number of people and teams involved in a decision leads to a fragmentation of responsibility. That is, rather than many people feeling responsible for a given outcome and thus making decisions with that end goal in mind, people can work to the specific goal of their work group, losing a sense of the bigger picture (Pidgeon & O'Leary, 2000). AI technologies amplify this issue of many hands, increasing the number of people involved in contributing to a decision, while also stretching the gap between actors and the resulting collective outcome. This phenomenon is articulated as moral distancing (Villegas-Galaviz & Martin, 2024) where developers (and in some contexts users) of AI are separated in various ways from those potentially impacted by the choices they make. A commonly cited example is the case of drones and remote warfare in which both developers and operators are not confronted with seeing the face of the opponent, making the decision to take life easier (Coeckelbergh, 2013; Cummings, 2004). Similar issues arise in less extreme contexts where harm may be an unintended consequence of the primary task of the technology.

AI creates multiple forms of distance. The first relates to proximity. Villegas-Galaviz and Martin (2024) articulate three forms of proximity distance: physical, temporal, and cultural. Physical distance refers to the geographic separation between the people and technology making decisions, and the location of their implementation. This has been raised in other contexts regarding the value of field experience for office-based engineers who, without such experience, may struggle to understand the size and scale of the technologies they are specifying and the resultant impact for workers in construction and operation (Maslen & Hayes, 2022). The lack of physical immediacy in the course of forming judgments can be exacerbated in the case of AI technologies which produce an analysis without "interacting with those who are affected by it" (Villegas-Galaviz & Martin, 2024, p. 1697). This physical distance has been observed by Montemayor et al. (2022) in a healthcare context, in which AI technologies support decision making at a distance meaning that decision makers (human or AI) tend to act without the empathy critical to clinical care.

There can also be significant elapsed time between a decision and the resultant outcome, as captured by the notion of temporal distance. AI models are necessarily developed well-ahead of their application. Once operational, the model changes with the introduction of new data too, meaning that it is "out of the hands of those who design, develop, and deploy AI," leaving an opening for unintended consequences (Villegas-Galaviz & Martin, 2024, p. 1698). Thus the very nature of AI means that even if an AI model is initially developed with the impact on users in mind, there is potential for it to learn over time in ways that move away from the initial intent. Another temporal factor relates to the long-term future impact of decisions. This is particularly a question of rare events as discussed in the next section.

The last form of proximity distance, which is cultural, refers to the gap between the context in which a model is developed and its potentially global application. While AI technologies may have the façade of being "neutral," they have the capacity to replicate and even amplify existing biases, as has been found in the cases of people of color (Benjamin, 2019; Noble, 2018), in relation to gender (O'Connor & Liu, 2024), and disability (Manzoor et al., 2025). These failures typically relate to issues of representation in training data (Paullada et al., 2021). While the literature on such biases is significant, in the context of technoscientific work distributed across multiple organizations, issues of organizational culture are also relevant. In Chap. 9, Jore and colleagues describe how the safety culture of offshore oil and gas workers is impacted by a new culture of security that has come about due to changes in national priorities being enacted organizationally. The significance of local cultures in global business is widely acknowledged in the organization studies literature (Ferraro & Briody, 2023), and yet there is a vacuum in consideration of this issue in relation to AI.

Related to issues of organizational culture, Villegas-Galaviz and Martin (2024) raise the problem of hierarchical distance, by which they mean the power differential at play within AI companies in which developers are pressured "to cut costs, increase profit and deliver higher quality. ... Managers might coerce ICT [Information and Communications Technology] professionals to make unethical or at least disputable decisions to the so-called benefit of the company" (Bergh & Deschoolmeester, 2010, pp. 1–2). This is a challenge for professionals beyond the AI context as we have seen throughout this volume illustrated perhaps most powerfully in the Titan case (Chap. 6) where professionals faced pressures from within their own organization to prioritize financial goals which were at odds with their ethical commitments as practicing engineers. It is noteworthy here that in many areas the generation of AI datasets, termed data annotating, is no longer treated as scientific work conducted within laboratories (Li, 2024) and is instead typically dispersed through gig platforms (Gray & Suri, 2019). The sheer volume of work involved in annotating large-scale datasets drove this shift, as Denton et al. (2021) traced in relation to ImageNet. While much of the sociological concern with this shift relates to precarity and new forms of work (Gray & Suri, 2019; Joyce et al., 2021; Shestakofsky, 2017), there is also a question around the implications of reclassifying these roles as unskilled labor rather than as specialized knowledge work undertaken by professionals whose actions are guided by codes of professional ethics. Jo and Gebru refer to the approach as laissez-faire in comparison to archival approaches, noting that AI data collection and annotation "is done without following a rigorous procedure or set of guidelines" (Jo & Gebru, 2020, p. 308). This hierarchical distance can also manifest in relations between the technology and users. We will discuss this further below in relation to the issue of over-reliance.

In considering these problems of distance, it is critical to understand that they cumulate and amplify each other to the point where no consideration is given to the ethics of technoscientific decisions. This has not been examined in relation to AI to the best of our knowledge, but the Grenfell Tower case discussed in Chap. 2 shows these multiple forms of distance at work. Building residents were far removed in location and time from the companies that manufactured the hazardous products used in the tower refurbishment. This made the ethics of building hazardous structures possible in the face of organizational cultures in the manufacturing companies that favored a deliberate focus on profit over regulatory compliance or concerns about safety.

Resolving the problem of distance is not straightforward, though the aspiration is that distance should be minimized. There are some efforts where the developers and the users closely collaborate in data annotation. In such instances, there is an observable focus on ethical matters. Zilberstein (2024) captured a case in which researchers working on AI in healthcare contexts seek to manage the ethical dilemmas involved in their work through drawing distinctions between that which is "real" in the clinic and the realm of the hypothetical that the AI dwells in through close collaboration with clinicians. Both the clinicians and the computer scientists are conscious of the documented potential harms that flow from the use of AI, such as how the technologies can reinforce biased stereotypes leading to different treatment for marginalized groups (Noble, 2018; Obermeyer et al., 2019; Zack et al., 2024). Vale (2024) presents a similar case in which research psychologists and psychiatrists navigate the ethics of digital psychiatry, which lacks ethical guidelines for dealing with patient training data. Vale describes these professionals as "moral entrepreneurs," who work to regulate the "Wild West" of digital psychiatry through processes of moral discovery and moral enclosure meaning articulation of a community of professionals who they recognize as "licensed to do the work of moral regulation" (Vale, 2024, p. 2).[1]

There is a great deal to be gleaned from developers and those involved in AI governance engaging closely with the work of professionals to deeply appreciate the nature of the work and its ethical challenges. Cases such as the ones that Zilberstein (2024) and Vale (2024) described are examples of bottom-up ethics, in that the professionals involved are working to form their own ethical standards, which in the healthcare case involves borrowing from established medical ethics. The involvement of users in the development of the training of models offers space to navigate the ethical questions raised. People working in the field who know well the ethical challenges that can emerge in the course of the work can witness emergent issues with AI outputs and are able to voice these ethical concerns. Involvement of the user group in the development of the tool is critical, but is also an idealized scenario considering corporate pressures. Nonetheless, this is a valuable approach.

3 Thinking About Rare Events

One of the challenges for technical professionals working in critical contexts is to identify and manage the potential for rare events. While technological disasters are reported every year, the ubiquity of technology in society means that a catastrophic failure in any given sociotechnical system as in the Boeing 737 MAX case described in Chap. 1 is a relatively rare

[1] These approaches have some relation to work in what is termed human-centered AI, where the focus is on reducing the distance between developers and those impacted by technology, i.e., the general public, rather than a professional group. For example, see Bingley et al. (2023).

event. Despite their relative rarity within any given sector or industry, such cases are memorable because of their impact. The potentially severe consequences mean that avoiding them is a key ethical issue, but their rarity makes them difficult to anticipate and so to prevent. This requires us to grapple with how experts working in these fields can see the latent faults and problems accumulating that mean a disaster is imminent and preventative action must be taken (Reason, 1997). With the step change in AI development, industries are understandably thinking about how AI might support these efforts, as well as the challenges of AI in this context.

As scholars working in this research area, seeing the potential for disaster is a problem that we have dedicated significant attention to throughout our research careers. Actions to prevent accidents arise at multiple system levels. Accidents are organizational failures and usually failures of regulation but professionals working with high stakes technologies have a role in preventing catastrophes too. The question then arises: How do professionals develop knowledge not only of what is probable but also of what is possible, especially in contexts in which people may never encounter the same feature or sequence of events twice in their career? Our focus in this work has been on supporting the knowledge and decision making capacity of human actors, and we have turned to stories as a proxy for experience, given the experience is both undesirable and unlikely (Hayes & Maslen, 2015; Maslen & Hayes, 2020). Stories can be a way to appreciate and share routine occurrences, but they are also powerful because of their capacity to capture outliers, i.e., those rare, seemingly minor problems in apparently unconnected parts of an organizational system that can create a fragility that is most easily seen in hindsight.

The intent in sharing these forms of knowledge through methods such as storytelling is to cultivate the ability of professionals to see a rising tide of problems that may have the potential for disaster and so give them the ability to raise the alarm and seek change before it is too late. When it comes to this kind of system performance monitoring, outliers in performance data, particularly across interlinked systems that may include both hardware and procedures, are the key "weak signals" (Nicolaidou et al., 2021) that experts seek out. Having such a sensitivity to operations is a key characteristic of a highly reliable organization (Weick et al., 1999). We were recently told by a major oil company that in 100% of failures in one division, individual professionals knew in advance that the system for which they were responsible was not operating well, but they assumed that the multiple other systems in place to prevent a system failure would act to ensure that no failure occurred. This false sense of security can only arise if no one is looking at the bigger picture.

It might seem that AI, with its ability to deal with big data, has the potential to identify developing system faults. In aviation, there has been a rapid increase in discussion of the potential for AI for safety, with a focus on using the technology to identify developing issues before they manifest (Morales et al., 2024). Key focus areas include weather but also monitoring of the behavior of front line personnel, e.g., the level of fatigue in operational staff including air traffic controllers and pilots (for a review, see Demir et al., 2024). But the standard way that AI models operate presents a problem. Regardless of their peculiarities, all models operate by looking for statistically significant patterns in data. They are ideally suited to identifying when system performance has drifted from what is desired but models can struggle to adequately account for all scenarios, especially rare cases, introducing safety concerns (Bai et al., 2025). In light of these known issues, researchers have mostly dedicated attention to methods that estimate the degree to which a model is accounting for rare events (for a summary of approaches, see Bai et al., 2025) rather than the implications of the weakness of the technology in accounting for rare events. Marsden and Steyer (2025, p. 13) pointed out that while AI technologies may support safety management, we might also find that "residual disasters will be more severe, as their early warning signals will less frequently be detected and managed in advance."

Another issue in AI accounting for multiple simultaneous system problems is the qualitative nature of some of the system performance data, particularly around human behaviors and actions. Development of an AI model involves "efforts to quantify the qualitative" (Monteiro, 2022, p. 169). Chapter 3 on how risk-based decisions are made in the gas pipeline sector illustrates vividly that expert judgment is a key, but often unacknowledged, part of making the best choices. Drawing an even wider net around system aspects that are useful performance indicators, Kok et al. (2020) advocate that "soft signals" are useful indicators of patient safety. They see that patient observations such as "the doctor was rude, the toilets are dirty" provide useful insights about the more specific experience of patient safety. Development of AI models will need to carefully consider the wide range of input data used in decisions of this kind. Models that do not (or perhaps cannot) consider all decision making inputs have the potential to create a mindfulness gap (Weick et al., 1999). This is subjective and the ability to see a problem depends on what one is open to seeing, as we saw in Chap. 7 where nurses and relatives could see there was a problem, but others could not. If a specific problem is not anticipated, no system of performance indicators is ever going to find it.

In addition to monitoring system performance, some AI proponents see a future where AI makes real-time decisions in operating technology. The issue of rare conditions also arises here. AI models for decision support are trained on datasets from actual operations in which abnormal operations by their very nature are rare. But human operators such as pilots are trained how to respond in an emergency situation. As one aviation sector expert explained: "I don't know how an AI agent will be supporting you because it's never seen one itself" (Hayes & Skogstad, 2025, p. 3095). In fact, AI outputs are constrained into a band of acceptable decisions for a variety of reasons, one of which being to guard against "hallucinations" (Pantha et al., 2024). Imagine, for instance, that aviation decision support tools were programmed to preclude the advice that the safest course of action was to land a passenger aircraft on a river following major engine failure. Pilots Chesley "Sully" Sullenberger and Jeffrey Skiles did this in what National Transportation Safety Board (NTSB) Board Member Kitty Higgins described as "the most successful ditching in aviation history" (Olshan, 2009) when US Airways Flight 1549 suffered a double engine failure as a result of a bird strike. Flight simulations following the failure suggested that the aircraft could have returned to LaGuardia or Teterboro Airport had it turned immediately (National Transportation Safety Board, 2010, pp. 49–50). However, the National Transportation Safety Board judged that this simulated decision course did not adequately account for real-world considerations "such as the time delay required to recognize the extent of the engine thrust loss and decide on a course of action" and that the ditching as executed by the pilots provided the highest probability of survival (National Transportation Safety Board, 2010, p. 89). It is difficult to see how such a course of action could be chosen by an AI in charge of an aircraft.

There are always valid exceptions to rules, and if rules are hard coded the possibility for those exceptions fall away. Time and again professionals in high stakes domains make incredible decisions such as the ones that Sullenberger and Skiles made that prevent catastrophe. Most of the time people beyond the professional network do not hear about them, because the failure was prevented before it occurred. Of course, there are times, too, in which the people in the room make a bad choice, distracted by political pressures as in the Challenger case (Vaughan, 1996), or commercial pressures as in the case of Deepwater Horizon (Hopkins, 2012). Pursuing decision support tools that have the potential to prevent these kinds of failures is a worthy goal. However, their deployment must be grounded in a robust understanding of the epistemic practices of professionals. As this discussion shows, in their current state, AI technologies are not at a point where they can operate in the service of preventing rare events.

4 Addressing Over-Reliance

Plans for use of AI in high stakes decision making in the short to medium term have assumed that final decision making authority and responsibility will lie with human actors. However, this design principle is on shaky foundations, perhaps reflective of wishful thinking on the part of developers and organizations. Humans, often, "will happily defer to the machine" (Zarsky, 2016) without knowing "how it works, what it does, and indeed what we are doing" (Coeckelbergh, 2013, p. 94). Gaube et al. (2021) conducted an experiment to examine how radiologists and physicians working in emergency departments respond to diagnostic advice that they know to be sourced from either humans or AI. This study did not find algorithmic aversion (Dietvorst et al., 2015) or algorithmic appreciation (Logg et al., 2019), as in other research. But what they did find is that people were generally likely to agree with advice, regardless of its source. This was especially the case for physicians who often look at x-rays as part of their role, but do not have the same task expertise as radiologists who specialize in this area. Over-reliance on AI could lead practitioners "to search for confirmatory information in place of conducting a thorough and critical evaluation" which is "dangerous … especially when the advice is inaccurate" (Gaube et al., 2021, pp. 4–5). This is a critical finding in how we understand the implications of introducing AI technologies into high stakes decision contexts more generally.

This issue is one of human cognition and requires specific consideration in the development and roll-out of AI technologies which are designed to do epistemic work in support of a human decision maker. Seeking advice from colleagues is a common decision making strategy, whether we are talking about medicine (Gaube et al., 2021), nuclear power generation (Hayes, 2013), or any number of other contexts. In particular, the study by Gaube and colleagues directs our attention to the timing of advice. People are more likely to discount the perspectives of others relative to their own, once they have formed their own judgment (Yaniv & Kleinberger, 2000). The problem of over-reliance could thus be countered if AI advice occurred following an initial assessment, or even only on request.

The difference between human and AI with respect to decision support also relates to the nature of the interaction. Seeking advice from a colleague typically follows a dialogical format where nuances and areas of uncertainty can be explored. Similar patterns are seen in group processes such as qualitative risk assessments as described in Chap. 3. In this case of high stakes decisions regarding public safety in the gas pipeline sector, expert consultants work to facilitate conversations regarding risk and risk controls and a significant part of their value is in forming the best engineering judgments through conversation. Dialogical approaches are particularly useful in navigating conflicting goals and areas of epistemic ambiguity (Maslen & Hayes, 2020). Use of AI recommendations as decision support typically does not follow this format. Gaube et al. (2021) suggest that physicians would benefit from interacting with AI technologies that communicate their degree of confidence and any limitations in their assessments, which is sound advice in relation to establishing epistemic trust but problematic given the lack of understanding of AI models at this point in time. In short, if we do not know how it works, we cannot interpret it as epistemically trustworthy (Alvarado, 2023).

This is not only an issue of human cognition. Work is often organized in such a way that professionals are under significant time pressure, and so advice on the right answer appears as a gift difficult to question as professionals run on with their days. We see this in the Grenfell Tower case (Chap. 2) where professionals working on the refurbishment of the building accepted without question the fraudulent fire safety certificates provided by building materials vendors, despite signs that regulatory compliance should be questioned more closely. Similar over-reliance on faulty data and a faulty model contributed to a natural gas pipeline rupture at San Bruno in California in 2010 (Hayes & Hopkins, 2014). When the pipeline was constructed in 1956, the sections of pipe that make up the pipeline included a longitudinal seam weld that was not properly made. This could have been discovered by pipeline inspection during the more than 50 years until the pipeline failed, but such inspections were never performed. Instead, company integrity engineers used a risk rating model to rank pipeline segments across their network to determine inspection priorities. The line that failed so catastrophically in 2010 was never inspected because of the ranking, which was based on flawed data. Indeed, the model outputs showed that the risk to the public from the pipeline network was actually decreasing. In short, the model was wrong, so its advice was wrong.

Over-reliance and its potential consequences are principally issues that need to be addressed at organizational and industry-wide levels. In high stakes decision making, aviation is the sector that is furthest advanced in working with AI technologies. While there is extensive conversation about the potential for AI in this domain, scholars and practitioners alike recognize the need for caution and slow adoption given the safety-critical nature of activities (Morales et al., 2024). AI-human partnering, rather than full automation or completely human operation, is viewed as the best way for the industry to achieve efficiency, customer service, and cost-effectiveness while also achieving safety goals (Cingöz & Durmaz, 2024). AI use in US and European aviation is currently restricted to systems that are not safety-critical and no Machine Learning is permitted during in flight operations (Hayes & Skogstad, 2025). AI technologies deployed on an aircraft, for example, therefore come with an integrated fixed database. More operational data may be gathered during each flight, but this can only be incorporated into models following development and recertification. There is broad understanding that further modes of use of AI must be allowed to reap further benefits that the technology presents, but this cannot be achieved with so many open questions of trust in the technology and ethics of its use.

In addition to being dangerous in the moment, over-reliance also has implications for the expertise development of professionals. If human actors are deferring to AI technologies, then they are likely orienting to the technology at the cost of developing the necessary knowledge to perform their roles independently and to take mindful and informed responsibility for outcomes. Often what is in focus is the new knowledge and skills required to work with AI, with multiple sectors calling for changes to degree programs with this in mind (e.g., see Gazquez-Garcia et al., 2025; Kabashkin et al., 2023). As the discussion of rare events brought into sharp focus, we need professionals to possess the necessary expertise to perform their roles beyond the skills needed to relate to decision support systems. Indeed, assessing the extent to which the advice from AI technologies (and other human actors for that matter) is sound requires a higher level of skill. The introduction of AI technologies is thus not only an issue of re-skilling, but preventing de-skilling.

5 Mechanisms for Raising Concerns

Another theme that authors have grappled with in this volume relates to questions of ethical regulation and governance (see Chaps. 8–11). Codes of practice and regulations are both forms of written rules that have two functions. Firstly, they attempt to provide proactive guidance for decision makers on what is socially acceptable. Secondly, they provide a minimum standard that can be referred to when things go wrong in the search for system improvements. Some governance arrangements make provisions for experts to speak up for ethical principles when they perceive that an organization is behaving badly. Whistleblowing provisions can be included in regulation covering technical requirements and are also typically a part of corporations law. In relation to AI, Papyshev and Chan (2025) warn of "fugazi" regulation, meaning something that is fake or phony, encompassing often vaguely defined rules that rely on high-level moral principles that are non-enforceable and do not engage with ethics in practice. The challenge of this "fugazi" regulation is that it does not offer clear guidelines for companies about what is, and what is not, acceptable. There are codes of practice in place for ICT professionals, though these relate to members of professional bodies as opposed to non-members and gig workers, with contention over the extent to which they make a measurable difference to ethical practice in this sector (Kluge Corrêa et al., 2025).

For professionals, this lack of clarity means that it is difficult to hold a "technical" line of defense, as we have heard some engineers call it (Maslen et al., 2021). Within their organizations, professionals often need to advocate for a given course of action on ethical grounds, whether or not they use that language. Having what can be hard conversations in the context of corporate pressures is greatly aided by an ability to point to externally mandated requirements (Maslen & Ransan-Cooper, 2017). While in the case of risk-based standards called up in regulation, these externally mandated requirements may still require the application of judgment and companies have different levels of maturity with respect to regulation (Parker et al., 2006), they nonetheless function as a critical resource. Chapter 10 on synthetic biology describes what can occur when there is a regulatory vacuum. The practice of "ethical licensing" has emerged where the patent holders of foundational technologies make licensing decisions based not only on commercial considerations but also on the ethics of the intended use. This leaves technology experts balancing multiple potentially conflicting values as they decide how their invention can be used. AI is not commercially restricted in the same way as the foundational knowledge is not so centralized, but without effective regulatory guidelines on use across a wide range of applications we could find ourselves in a similar situation of computer scientists deciding what is right without the necessary subject matter expertise to make such judgments.

Where these internal interventions fail, whistleblowing requires clear regulation and, especially relevant in the case of AI, a clear understanding of how the technologies operate, for it to be possible for professionals to articulate the nature of ethical concerns. AI whistleblowers may be developers working in private companies or users such as medical professionals and engineers. There have been some high-profile cases in recent years of insiders, particularly with private technology companies, raising concerns. For instance, OpenAI employees wrote an open letter, raising that "AI companies have strong financial incentives to avoid effective oversight" and that "employees are among the few people who can hold them accountable" (Roose, 2024). Wu (2024, p. 2) argues that the significant financial pressures coupled with escalating expectations over what AI technologies can achieve "may incentivize risk-taking behaviors" on the part of companies "making whistleblowers increasingly essential to expose shortcuts, ethical lapses, or governance failures before they cause harm." These pressures that encourage risk taking coupled with the consequences of failure underscore the importance of clear regulation (Marsden & Steyer, 2025).

A key reason that whistleblowers are critical in relation to AI, perhaps more so than in other domains, relates to informational asymmetry. The computer scientists working inside companies have the expertise to identify cases of wrongdoing accurately (Olesen, 2025). AI auditing has been proposed as a governance mechanism (for a review, see Mökander, 2023), but it can be difficult to assess whether AI technologies are meeting the required ethical standards, with scholars investigating different methods for a robust outsider assessment of AI, such as the use of large interdisciplinary teams including domain experts, technical experts, and ethicists, among others (e.g., see Vetter et al., 2023). Even with this level of labor dedicated to an auditing exercise, it can be the case that such external teams can fail to identify problems that are known to people who develop and work with the technology (Wu, 2024), an issue that has been observed in relation to technical regulation more generally (Hopkins, 2007; King & Hayes, 2018). There is also an issue of proximity. Auditing is a moment in time, where whistleblowers have a "front row seat" (Olesen, 2025).

While important, AI whistleblowing is challenging for a variety of reasons. Given the role of insiders in calling out ethical concerns, commentators have dedicated attention to whistleblower regimes, noting that, at least under US state and federal law, whistleblower protections are ambiguous due to the largely unregulated nature of AI development (Wu, 2024). AI development is increasingly taking place in highly secretive environments due to national security and cybersecurity concerns; nondisclosure agreements are prevalent due to corporate secrecy policies. The nature of the threat is also typically anticipated in the future, or characterized by uncertainty, rather than an immediate threat and/or law violation (Wu, 2024). The Titan case (Chap. 6) where complaints were treated as an employment dispute and the Gosport case (Chap. 7) where reports of problems were dismissed illustrate vividly how difficult it can be for whistleblowers to initiate change. Unclear protections, the development environments, and the nature of the threat amplify these difficulties.

6 A Final Note on Responsibility

A significant question in the deployment of AI technologies into technoscientific domains relates to responsibility. Put simply, who is responsible when things go wrong where decision making is distributed between human and nonhuman actors? This is an issue that is typically poorly considered in the roll-out of AI technologies, leaving professionals in what Holford (2022, p. 141) described as a "legal crumple zone." Responsibility must be front and center in the development and deployment of AI technologies, hopefully not for the purposes of assigning blame, but to heighten consideration of epistemic trust, the capacity of AI for ethical reasoning, and the relations between human and nonhuman.

Carboni et al. (2023) have examined these issues in relation to AI-assisted histopathology where, organizationally, it appears as if the AI is making the decision, with the human actors effectively hidden. The authors emphasize that this is a case of "fauxtomation" given the significant work required on the part of human attendants including manually focusing on aspects of the tissue based on instructions from the pathologist. In doing this focusing work the administrative assistant who scans the slides "lend[s] her eyes and her contextualized seeing to the scanner" with this work essential to the production of "good digital images" (Carboni et al., 2023, p. 11). Pathologists working as part of these teams also rarely base their findings on the digital slides alone, moving instead back to the physical sample as viewable under a microscope, suggesting that the digitizing process changes what can be seen and so the certainty with which pathologists can offer a diagnosis. One pathologist described the difference in relation to seeing a granulocyte (a type of white blood cell), where under the microscope she can see "granules," which on the digital slide appear as red dots. Knowing this difference in presentation she could adjust her assessment to read the red dots as the diagnostically significant granules, though she noted that other forms of information may be missing entirely (Carboni et al., 2023, p. 14).

This case serves as a check on AI imaginaries that overblow the degree to which AI replaces the role of human judgment. More than this though, this arrangement requires us to think seriously about who, or what, is the ultimate decision maker. Studies of adoption of AI diagnostic technologies in the healthcare context repeatedly show that trust, and so acceptance of the technology, increased where patients had a sense that their doctors were the ones taking the decision, rather than deferring to AI (Hallowell et al., 2023; Robertson et al., 2023), a perspective shared among healthcare professionals themselves (Sarwar et al., 2019). Cases where human actors are hidden, as Carboni et al. (2023) describe, raise questions about what professionals should do when they see something that the AI has not.

The aviation sector is again leading conversations on AI responsibility. Holford (2022) cautions against shared responsibility, arguing that pilots should retain skills to give them overall mastery and therefore both control and responsibility for outcomes. To consider AI as more than an assistant requires deep consideration of AI failure modes "which may be subtle and hard or even impossible to detect" (Kirwan, 2025). This must logically lead to limits on the uses of AI technologies, noting the problems with over-reliance that we described earlier. Even if we assume that human actors are able to retain their own judgments of a situation or problem, AI technologies themselves are typically inscrutable, and so the human ability to interrogate the AI analysis is limited (Martin, 2019).

As we have already acknowledged, the specifics of given AI technologies including their scope, sector, and the nature of decision making (i.e., are decisions high stakes) are vital to keep in focus. To support best practice in the development, roll-out, and use of AI technologies, we have thus compiled a series of reflective questions rather than seeking to make a generalized conclusion for the management of AI as an epistemic technology. The first set of questions is geared toward considerations in the design of AI technologies and their implications for the organization of work. The questions in the second set are focused on the use of AI technologies in decision support, picking up the considerations related to responsibility, epistemic trust, and the nature of the human/AI relation. The final set of questions relates to matters of regulation and governance, directing consideration toward a thorough engagement with ethics in practice including the development of good rules of thumb and pathways for bad news to travel. We hope that these are productive for scholars, as well as people in practice, as we work to understand more about the pitfalls and possibilities of thinking with AI (Table 12.1).

Table 12.1 Reflective questions for development and use of AI technologies in decision support in critical contexts

Sector and organizational readiness	Are subject matter professionals involved in all stages of the development of the AI technologies? How are professional ethics informing the design of AI technologies? What are the potential consequences of decisions made? Can the AI account for rare events? How is professionals' experience with use of the AI technologies fed back into the design of the technology itself? Are professionals trained in decision making in the context of AI technologies? How does the introduction of AI technologies change work roles within the organization?
Knowledge and decision making	What is the epistemic role of AI within the decision making process? How much autonomy does the AI have? Do professionals have an ability to understand the basis of the AI output, including how this may change over time? What is the timing of the AI input to decision making relative to initial professional judgment? Is there capacity for dialogical reasoning between the human and AI? Is the AI output trustworthy? That is, are there known cases of "hallucinations"? How would professionals being advised by the AI know if the AI has made an error? Will a human actor be able to intervene in cases of identified AI errors? How will the AI respond if it judges that a human has made a mistake? How will observed ethical conflicts between professionals and AI technologies be managed?
Governance	What professional codes of practice/ethics are in place to support professionals in the context of AI? Are professional codes of practice/ethics sufficiently grounded in the practical considerations and challenges for a given sector? Are there clear regulations in place for AI technologies in the sector? Who is ultimately responsible for a decision made with AI? Is this clear to everyone involved? What are the ethical obligations of professionals and organizations? What communication channels are in place for professionals to raise concerns over and above normal operational feedback?

References

Algera, E. (2023). Knowing (with) the body: Sensory knowing in contraceptive self-tracking. *Sociology of Health & Illness, 45*(2), 242–258. https://doi.org/10.1111/1467-9566.13570

Alvarado, R. (2023). AI as an epistemic technology. *Science and Engineering Ethics, 29*(5), Article 32. https://doi.org/10.1007/s11948-023-00451-3.

Asif, S., Wenhui, Y., ur-Rehman, S., ul-ain, Q., Amjad, K., Yueyang, Y., Jinhai, S., & Awais, M. (2025). Advancements and prospects of machine learning in medical diagnostics: Unveiling the future of diagnostic precision. *Archives of Computational Methods in Engineering, 32*(2), 853–883. https://doi.org/10.1007/s11831-024-10148-w

Australian Medical Association. (2016). *AMA Code of Ethics.* https://www.ama.com.au/sites/default/files/2021-02/AMA_Code_of_Ethics_2004._Editorially_Revised_2006._Revised_2016_0.pdf

Bai, Y.-L., Huang, Z.-Y., Lam, H., & Zhao, D. (2025). Black-box rare-event simulation for safety testing of AI agents: An overview. *Journal of the Operations Research Society of China, 13*(3), 750–774. https://doi.org/10.1007/s40305-025-00585-0

Benjamin, R. (2019). *Race after technology: Abolitionist tools for the new Jim code*. Polity.

Bergh, J. V. d., & Deschoolmeester, D. A. K. (2010). Ethical decision making in ICT: Discussing the impact of an ethical code of conduct. *Communications of The IbIMA, 2010*, 1–11. https://doi.org/10.5171/2010.127497

Bingley, W. J., Curtis, C., Lockey, S., Bialkowski, A., Gillespie, N., Haslam, S. A., Ko, R. K. L., Steffens, N., Wiles, J., & Worthy, P. (2023). Where is the human in human-centered AI? Insights from developer priorities and user experiences. *Computers in Human Behavior, 141*, 107617. https://doi.org/10.1016/j.chb.2022.107617

Buccella, A. (2025). Ethically charged decisions and the future of 'AI ethics'. *AI & SOCIETY*. https://doi.org/10.1007/s00146-025-02573-7

Carboni, C., Wehrens, R., van der Veen, R., & de Bont, A. (2023). Eye for an AI: More-than-seeing, fauxtomation, and the enactment of uncertain data in digital pathology. *Social Studies of Science, online first, 03063127231167589.* https://doi.org/10.1177/03063127231167589

Cingöz, K. N., & Durmaz, V. (2024). Beyond the horizon: Exploring AI'S role in enhancing aviation safety and efficiency—An airline perspective. In S. Kurnaz, A. Rodrigues, & A. Padhra (Eds.), *Strategic management and policy in the global aviation industry* (pp. 203–222). IGI Global Scientific Publishing. https://doi.org/10.4018/979-8-3693-0908-7.ch011

Coeckelbergh, M. (2013). Drones, information technology, and distance: Mapping the moral epistemology of remote fighting. *Ethics and Information Technology, 15*(2), 87–98. https://doi.org/10.1007/s10676-013-9313-6

Coeckelbergh, M. (2020). *AI ethics*. The MIT Press.

Cummings, M. (2004). Creating moral buffers in weapon control interface design. *IEEE Technology and Society Magazine, 23*(3), 28–33. https://doi.org/10.1109/MTAS.2004.1337888

Dafoe, A. (2018). AI governance: A research agenda. https://cdn.governance.ai/GovAI-Research-Agenda.pdf.

Dahlin, E. (2024). Think differently we must! An AI manifesto for the future. *AI & SOCIETY, 39*(3), 1423–1426. https://doi.org/10.1007/s00146-022-01620-x

Demir, G., Moslem, S., & Duleba, S. (2024). Artificial intelligence in aviation safety: Systematic review and biometric analysis. *International Journal of Computational Intelligence Systems, 17*(1), Article 279. https://doi.org/10.1007/s44196-024-00671-w

Denton, E., Hanna, A., Amironesei, R., Smart, A., & Nicole, H. (2021). On the genealogy of machine learning datasets: A critical history of ImageNet. *Big Data & Society, 8*(2), 20539517211035955. https://doi.org/10.1177/20539517211035955

Dietvorst, B. J., Simmons, J. P., & Massey, C. (2015). Algorithm aversion: People erroneously avoid algorithms after seeing them err. *Journal of Experimental Psychology: General, 144*(1), 114–126. https://doi.org/10.1037/xge0000033

Elyan, E., Jamieson, L., & Ali-Gombe, A. (2020). Deep learning for symbols detection and classification in engineering drawings. *Neural Networks, 129*, 91–102. https://doi.org/10.1016/j.neunet.2020.05.025

Engineers Australia. 2022. Code of Ethics and Guidelines on Professional Conduct. edited by E. Australia. Canberra: Engineers Australia.

Ferraro, G., & Briody, E. (2023). *The cultural dimension of global business* (9th ed.). Routledge. https://doi.org/10.4324/9781003214144

Floridi, L., & Cowls, J. (2022). A unified framework of five principles for AI in society. In *Machine learning and the city* (pp. 535–545). https://doi.org/10.1002/9781119815075.ch45.

Fotopoulou, A., & O'Riordan, K. (2017). Training to self-care: Fitness tracking, biopedagogy and the healthy consumer. *Health Sociology Review, 26*(1), 54–68. https://doi.org/10.1080/14461242.2016.1184582

Gaube, S., Suresh, H., Raue, M., Merritt, A., Berkowitz, S. J., Lermer, E., Coughlin, J. F., Guttag, J. V., Colak, E., & Ghassemi, M. (2021). Do as AI say: Susceptibility in deployment of clinical decision-aids. npj Digital Medicine, *4*(1), Article 31. https://doi.org/10.1038/s41746-021-00385-9.

Gazquez-Garcia, J., Sánchez-Bocanegra, C. L., & Sevillano, J. L. (2025). AI in the health sector: Systematic review of key skills for future health professionals. *JMIR Med Educ, 11*, e58161. https://doi.org/10.2196/58161

Goodfellow, I., Bengio, Y., & Courville, A. (2016). *Deep learning*. The MIT Press.

Gray, M., & Suri, S. (2019). *Ghost work: How to stop Silicon Valley from building a new global underclass*. Houghton Millfin Harcourt.

Green, B. (2021). The contestation of tech ethics: A sociotechnical approach to technology ethics in practice. *Journal of Social Computing, 2*(3), 209–225. https://doi.org/10.23919/JSC.2021.0018

Hallowell, N., Badger, S., McKay, F., Kerasidou, A., & Nellåker, C. (2023). Democratising or disrupting diagnosis? Ethical issues raised by the use of AI tools for rare disease diagnosis. *SSM—Qualitative Research in Health, 3*, 100240. https://doi.org/10.1016/j.ssmqr.2023.100240

Harris, A. (2016). Listening-touch, affect and the crafting of medical bodies through percussion. *Body & Society, 22*(1), 31–61. https://doi.org/10.1177/1357034X15604031

Hauswald, R. (2025). AI and the philosophy of expertise and epistemic authority. In M. Hähnel & R. Müller (Eds.), *A companion to applied philosophy of AI* (pp. 55–70). https://doi.org/10.1002/9781394238651.ch5.

Hayes, J. (2013). *Operational decision-making in high-hazard organizations: Drawing a line in the sand*. Ashgate.

Hayes, J., & Hopkins, A. (2014). *Nightmare pipeline failures: Fantasy planning, black swans and integrity management*. CCH Australia.

Hayes, J., & Maslen, S. (2015). Knowing stories that matter: Learning for effective safety decision-making. *Journal of Risk Research, 18*(6), 714–726. https://doi.org/10.1080/13669877.2014.910690

Hayes, J., & Skogstad, M. R. (2025). From rigid orderliness to barely controlled chaos: sociotechnical risk and AI in aviation. In *Proceedings of the 35th European Safety and Reliability Conference and the 33th Society for Risk Analysis Europe Conference* (Stavanger, Norway).

Holford, W. D. (2022). An ethical inquiry of the effect of cockpit automation on the responsibilities of airline pilots: Dissonance or meaningful control? *Journal of Business Ethics, 176*(1), 141–157. https://doi.org/10.1007/s10551-020-04640-z

Hopkins, A. (2007). Beyond compliance monitoring: New strategies for safety regulators. *Law & Policy, 29*(2), 210–225. https://doi.org/10.1111/j.1467-9930.2007.00253.x

Hopkins, A. (2012). *Disastrous decisions: The human causes of the Gulf of Mexico blowout*. CCH Australia.

Huang, L., Yu, W., Ma, W., Zhong, W., Feng, Z., Wang, H., Chen, Q., Peng, W., Feng, X., Qin, B., & Liu, T. (2025). A survey on hallucination in large language models: Principles, taxonomy, challenges, and open questions. *ACM*

Transactions on Information Systems, 43(2), Article 42. https://doi.org/10.1145/3703155.

Islam, T., Hafiz, M. S., Jim, J. R., Kabir, M. M., & Mridha, M. F. (2024). A systematic review of deep learning data augmentation in medical imaging: Recent advances and future research directions. *Healthcare Analytics, 5*, 100340. https://doi.org/10.1016/j.health.2024.100340

Janiesch, C., Zschech, P., & Heinrich, K. (2021). Machine learning and deep learning. *Electronic Markets, 31*(3), 685–695. https://doi.org/10.1007/s12525-021-00475-2

Jiao, Z., Hu, P., Xu, H., & Wang, Q. (2020). Machine learning and deep learning in chemical health and safety: A systematic review of techniques and applications. *ACS Chemical Health & Safety, 27*(6), 316–334. https://doi.org/10.1021/acs.chas.0c00075

Jo, E. S., & Gebru, T. (2020). Lessons from archives: Strategies for collecting sociocultural data in machine learning. In *Proceedings of the 2020 conference on fairness, accountability, and transparency* (pp. 306–316). https://doi.org/10.1145/3351095.3372829.

Joyce, K., Smith-Doerr, L., Alegria, S., Bell, S., Cruz, T., Hoffman, S. G., Noble, S. U., & Shestakofsky, B. (2021). Toward a sociology of artificial intelligence: A call for research on inequalities and structural change. *Socius, 7*, 2378023121999581. https://doi.org/10.1177/2378023121999581

Kabashkin, I., Misnevs, B., & Zervina, O. (2023). Artificial intelligence in aviation: New professionals for new technologies. *Applied Sciences, 13*(21), 11660. https://www.mdpi.com/2076-3417/13/21/11660

King, D. K., & Hayes, J. (2018, 2018/09/02). The effects of power relationships: Knowledge, practice and a new form of regulatory capture*. Journal of Risk Research, 21(9), 1104–1116. https://doi.org/10.1080/13669877.2017.1382560.

Kirwan, B. (2025). Human factors requirements for human-AI teaming in aviation. *Future Transportation, 5*(2), 42. https://www.mdpi.com/2673-7590/5/2/42

Kluge Corrêa, N., William Santos, J., Almeida, É. d., Pasetti, M., Schiavon, D., Panizzon, M., & Oliveira, N. d. (2025). Codes of ethics in IT: Do they work in isolation? *AI and Ethics, 5*(4), 4337–4349. https://doi.org/10.1007/s43681-025-00744-2

Kok, J., Wallenburg, I., Leistikow, I., & Bal, R. (2020). "The doctor was rude, the toilets are dirty": Utilizing 'soft signals' in the regulation of patient safety. *Safety Science, 131*, 104914. https://doi.org/10.1016/j.ssci.2020.104914

Kwon, Y., & Hong, S. (2025). The fall and rise of Iruda: Reassembling AI through ethics-in-action. *Social Studies of Science*, online first https://doi.org/10.1177/03063127251360397.

Li, Z. (2024). When being a data annotator was not yet a job: The laboratory origins of dispersible labor in computer vision research. *Socius, 10*, 23780231241259617. https://doi.org/10.1177/23780231241259617

Logg, J. M., Minson, J. A., & Moore, D. A. (2019). Algorithm appreciation: People prefer algorithmic to human judgment. *Organizational Behavior and Human Decision Processes, 151*, 90–103. https://doi.org/10.1016/j.obhdp.2018.12.005

Lupton, D., & Maslen, S. (2018). The more-than-human sensorium: Sensory engagements with digital self-tracking technologies. *The Senses & Society, 13*(2), 190–202. https://doi.org/10.1080/17458927.2018.1480177

Manzoor, R., Hussain, W., & Anjum, M. L. (2025). Out of dataset, out of algorithm, out of mind: A critical evaluation of AI bias against disabled people. *AI & SOCIETY, 40*(5), 3941–3951. https://doi.org/10.1007/s00146-024-02168-8

Marsden, E., & Steyer, V. (2025). *Artificial intelligence and safety management: an overview of key challenges* (FonCSI Cahier No. 2025–03). The Foundation for an Industrial Safety Culture. https://doi.org/10.57071/iae289.

Martin, K. (2019). Ethical implications and accountability of algorithms. *Journal of Business Ethics, 160*(4), 835–850. https://doi.org/10.1007/s10551-018-3921-3

Maslen, S. (2025). *Learning to hear: The auditory bases of excellence in practicing medicine, climbing mountains, making music, and communicating in Morse code*. Columbia University Press.

Maslen, S., & Harris, A. (2021). Becoming a diagnostic agent: A collated ethnography of digital-sensory work in caregiving intra-actions. *Social Science & Medicine, 277*, 113927. https://doi.org/10.1016/j.socscimed.2021.113927

Maslen, S., & Hayes, J. (2020). "This is how we debate": Engineers' use of stories to reason through disaster causation. *Qualitative Sociology, 43*(2), 191–212. https://doi.org/10.1007/s11133-020-09452-1

Maslen, S., & Hayes, J. (2022). "It's the seeing and feeling": How embodied and conceptual knowledges relate in pipeline engineering work. *Qualitative Sociology, 45*(4), 593–616. https://doi.org/10.1007/s11133-022-09520-8

Maslen, S., Hayes, J., Holdsworth, S., & Sandri, O. (2021). When ethics is a technical matter: Engineers' strategic appeal to ethical considerations in advocating for system integrity. *Science and Engineering Ethics, 27*(4), 46. https://doi.org/10.1007/s11948-021-00324-7

Maslen, S., & Ransan-Cooper, H. (2017). Safety framing and compliance in relation to standards: Experience from the Australian gas pipeline industry. *Safety Science, 94*, 52–60. https://doi.org/10.1016/j.ssci.2016.12.011

Mökander, J. (2023). Auditing of AI: Legal, ethical and technical approaches. *Digital Society, 2*(3), article 49. https://doi.org/10.1007/s44206-023-00074-y.

Monteiro, E. (2022). *Digital oil: Machineries of knowing*. The MIT Press. https://doi.org/10.7551/mitpress/14604.001.0001

Montemayor, C., Halpern, J., & Fairweather, A. (2022). In principle obstacles for empathic AI: Why we can't replace human empathy in healthcare. *AI & SOCIETY, 37*(4), 1353–1359. https://doi.org/10.1007/s00146-021-01230-z

Morales, V. E., Sanchez, J., Escalera, J., Sharma, V., & Wheeler, B. E. (2024). Artificial intelligence and aviation: Content analysis of research publications from 2013–2023. *The Journal of Management and Engineering Integration, 17*(2). https://doi.org/10.62704/10057/28469

Myers, N. (2008). Molecular embodiments and the body-work of modeling in protein crystallography. *Social Studies of Science, 38*(2), 163–199. https://doi.org/10.1177/0306312707082969

National Transportation Safety Board. (2010). *Aircraft accident report NTSB/AAR-10/03: Loss of thrust in both engines after encountering a flock of birds and subsequent ditching on the Hudson River, US Airways Flight 1549 Airbus A320–214 N106US, Weehawken, New Jersey, January 15, 2009*. https://www.ntsb.gov/investigations/AccidentReports/Reports/AAR1003.pdf

Nicolaidou, O., Dimopoulos, C., Varianou-Mikellidou, C., Boustras, G., & Mikellides, N. (2021). The use of weak signals in occupational safety and health: An investigation. *Safety Science, 139*, 105253. https://doi.org/10.1016/j.ssci.2021.105253

Noble, S. U. (2018). *Algorithms of oppression: How search engines reinforce racism*. NYU Press.

O'Connor, S., & Liu, H. (2024). Gender bias perpetuation and mitigation in AI technologies: Challenges and opportunities. *AI & SOCIETY, 39*(4), 2045–2057. https://doi.org/10.1007/s00146-023-01675-4

Obermeyer, Z., Powers, B., Vogeli, C., & Mullainathan, S. (2019). Dissecting racial bias in an algorithm used to manage the health of populations. *Science, 366*(6464), 447–453. https://doi.org/10.1126/science.aax2342

Olesen, T. (2025). Big tech whistleblowing: Frances Haugen and the Facebook files. *Organization*, Advance online publication. https://doi.org/10.1177/13505084251321785.

Olshan, J. (2009, January 17). Quiet air hero is Captain America. *New York Post*. https://nypost.com/2009/01/17/quiet-air-hero-is-captain-america/

Oudshoorn, N. (2008). Diagnosis at a distance: The invisible work of patients and healthcare professionals in cardiac telemonitoring technology. *Sociology of Health & Illness, 30*(2), 272–288. https://doi.org/10.1111/j.1467-9566.2007.01032.x

Pantha, N., Ramasubramanian, M., Gurung, I., Maskey, M., & Ramachandran, R. (2024). Challenges in guardrailing large language models for science. *arXiv*. https://arxiv.org/abs/2411.08181

Papyshev, G., & Chan, K. J. D. (2025). Fugazi regulation for AI: Strategic tolerance for ethics washing. *AI & SOCIETY, 40*(5), 3475–3486. https://doi.org/10.1007/s00146-024-02084-x

Parker, D., Lawrie, M., & Hudson, P. (2006). A framework for understanding the development of organisational safety culture. *Safety Science, 44*(6), 551–562. https://doi.org/10.1016/j.ssci.2005.10.004

Paullada, A., Raji, I. D., Bender, E. M., Denton, E., & Hanna, A. (2021). Data and its (dis)contents: A survey of dataset development and use in machine learning research. *Patterns, 2*(11), 100336. https://doi.org/10.1016/j.patter.2021.100336

Pidgeon, N., & O'Leary, M. (2000). Man-made disasters: Why technology and organizations (sometimes) fail. *Safety Science, 34*(1–3), 15–30. https://doi.org/10.1016/S0925-7535(00)00004-7

Pink, S., Sumartojo, S., Lupton, D., & Heyes La Bond, C. (2017). Mundane data: The routines, contingencies and accomplishments of digital living. *Big Data & Society, 4*(1), 2053951717700924. https://doi.org/10.1177/2053951717700924

Prentice, R. (2013). *Bodies in formation: An ethnography of anatomy and surgery education*. Duke University Press.

Reason, J. (1997). *Managing the risks of organizational accidents*. Ashgate.

Robertson, C., Woods, A., Bergstrand, K., Findley, J., Balser, C., & Slepian, M. J. (2023). Diverse patients' attitudes towards artificial intelligence (AI) in diagnosis. *PLOS Digital Health, 2*(5), e0000237. https://doi.org/10.1371/journal.pdig.0000237

Roose, K. (2024). OpenAI insiders warn of a 'reckless' race for dominance. *New York Times*. https://www.nytimes.com/2024/06/04/technology/openai-culture

Sarwar, S., Dent, A., Faust, K., Richer, M., Djuric, U., Van Ommeren, R., & Diamandis, P. (2019). Physician perspectives on integration of artificial intelligence into diagnostic pathology. *npj Digital Medicine, 2*(1), 28. https://doi.org/10.1038/s41746-019-0106-0

Shestakofsky, B. (2017). Working algorithms: Software automation and the future of work. *Work and Occupations, 44*(4), 376–423. https://doi.org/10.1177/0730888417726119

Shin, D. (2025). Can AI have a sense of morality? *AI & SOCIETY, 40*(6), 4169–4170. https://doi.org/10.1007/s00146-025-02476-7

Sterne, J. (2001). Mediate auscultation, the stethoscope, and the "autopsy of the living": Medicine's acoustic culture. *Journal of Medical Humanities, 22*(2), 115–136. https://doi.org/10.1023/a:1009067628620

Thompson, D. (1980). Moral responsibility of public officials: The problem of many hands. *American Political Science Review, 74*(4), 905–916. https://doi.org/10.2307/1954312

Vale, M. D. (2024). Moral entrepreneurship and the ethics of artificial intelligence in digital psychiatry. *Socius, 10*, 23780231241259641. https://doi.org/10.1177/23780231241259641

van de Poel, I., Royakkers, L., & Zwart, S. D. (2015). *Moral responsibility and the problem of many hands.* Routledge.

Vaughan, D. (1996). *The Challenger launch decision: Risky technology, culture and deviance at NASA.* University of Chicago Press.

Vertesi, J. (2015). *Seeing like a rover: How robots, teams, and images craft knowledge of Mars.* University of Chicago Press.

Vetter, D., Amann, J., Bruneault, F., Coffee, M., Düdder, B., Gallucci, A., Gilbert, T. K., Hagendorff, T., van Halem, I., Hickman, E., Hildt, E., Holm, S., Kararigas, G., Kringen, P., Madai, V. I., Wiinblad Mathez, E., Tithi, J. J., Westerlund, M., Wurth, R., Zicari, R. V., & Initiative, Z. I. (2023). Lessons learned from assessing trustworthy AI in practice. *Digital Society, 2*(3), 35. https://doi.org/10.1007/s44206-023-00063-1

Villegas-Galaviz, C., & Martin, K. (2024). Moral distance, AI, and the ethics of care. *AI & SOCIETY, 39*(4), 1695–1706. https://doi.org/10.1007/s00146-023-01642-z

Weick, K., Sutcliffe, K., & Obstfeld, D. (1999). Organizing for high reliability: Processes of collective mindfulness. In R. I. Sutton & B. M. Staw (Eds.), *Research in organizational behavior* (pp. 81–123). JAI Press.

Wu, H. (2024). AI whistleblowers. *SSRN.* https://doi.org/10.2139/ssrn.4790511

Yaniv, I., & Kleinberger, E. (2000). Advice taking in decision making: Egocentric discounting and reputation formation. *Organizational Behavior and Human Decision Processes, 83*(2), 260–281. https://doi.org/10.1006/obhd.2000.2909

Zack, T., Lehman, E., Suzgun, M., Rodriguez, J. A., Celi, L. A., Gichoya, J., Jurafsky, D., Szolovits, P., Bates, D. W., Abdulnour, R.-E. E., Butte, A. J., & Alsentzer, E. (2024). Assessing the potential of GPT-4 to perpetuate racial and gender biases in health care: A model evaluation study. *The Lancet Digital Health, 6*(1), e12–e22. https://doi.org/10.1016/S2589-7500(23)00225-X

Zarsky, T. (2016). The trouble with algorithmic decisions: An analytic road map to examine efficiency and fairness in automated and opaque decision making. *Science, Technology, & Human Values, 41*(1), 118–132. https://doi.org/10.1177/0162243915605575

Zilberstein, S. (2024). Ethical dilemmas and collaborative resolutions in machine learning research for health care. *Socius, 10*, 23780231241259671. https://doi.org/10.1177/23780231241259671

Index

A
Administrative evil, 29–32, 40, 44
Air traffic control, 260, 267
Ambiguity, 151, 179, 254, 270
Artificial intelligence (AI), 4, 13, 17, 20, 21, 93, 105, 107
As low as reasonably practicable (ALARP), 49–65, 201
Aviation, 1–4, 267, 268, 271, 275

B
Bad apples, 29, 43
Belonging, professional, 164, 176, 250–252
Bentham, Jeremy, 54
Biomedical education, 12
Broad Institute, 214, 216, 217, 221–228, 230–232
Built environment, 11, 264

C
Carson, Rachel, 69
Certification, *see* Testing and certification
Citizen activism, 12, 70
Civil society, 70, 82, 85–88, 208, 260
Climate change, 5, 8, 17
Codes of ethics, 4, 6, 9, 15–17, 19, 20, 167, 168, 178, 180, 261, 272, 276
Collections of human origin, 12, 91–111
Colonialism, 171, 172
Complaints, *see* Speaking up
Compliance
  regulatory, 9, 15, 18, 19, 40, 42, 62, 64, 88, 119, 124, 205, 231, 240, 264, 270
  technical, 43, 241
Concerns, *see* Speaking up
Confucian ethics, 71
Conscience, 6, 58
Consent, 12, 91–111, 151, 207, 209, 222, 223, 239, 247, 251

J. Hayes, S. Maslen (eds.), *Ethics in the doing of technoscience*,
https://doi.org/10.1007/978-3-032-20002-0

CRISPR, 19, 213–233
Culture
  community, 100–102, 109, 110
  engineering, 11, 18, 163–167, 169, 172, 177, 179
  learning, 119
  medical, 157, 158
  organizational, 2, 8, 31, 119, 124, 166, 177, 263, 264
  professional, 31, 168
  safety, 18, 116, 263
  workplace, 202, 240

D

Decision making, 7, 8, 21, 28, 63, 64, 96, 104, 105, 108, 142, 160, 166, 167, 219, 239
  ambiguity, 168
  authority, 125
  communication, 116, 118, 131, 151
  consensus, 57
  consequences of, 118, 151
  cost benefit analysis, 50, 53, 54, 60–61, 72
  framing, 43
  high stakes, 8, 49–65, 115–137, 187–209
  informed, 12, 86
  judgment, 6, 7, 10, 11, 20, 57–59, 62, 117, 118, 262, 267, 270, 272, 273, 275, 276
  justification, 61
  management, 8, 246–249, 254
  moment, 21, 261
  occupational safety and health, 238
  power and, 115–137
  processes, 158
  qualitative methods, 59, 60
  quantitative methods, 59
  risk-based, 63
  shared, 208
  tradeoffs, 44, 123
  values-based, 133
Deep Learning, 107, 260
Diversity, equity, and inclusion (DEI), *see* Diversity
Deontology, 5, 9
Dignity, 12, 99, 101–105, 108–111, 151, 250
Dilemmas, ethical, 7, 18, 19, 101, 132, 168–170, 172, 174–176, 179, 180, 189, 193, 198, 207, 209, 238, 264
Disaster cases
  Bhopal, 32
  Boeing 737 MAX, 1–4, 116, 265
  Challenger space shuttle loss, 116, 268
  Columbia space shuttle loss, 116
  CRISPR babies, 213, 215, 217, 222
  Deepwater Horizon blowout, 116, 269
  Dubai cladding fires, 35
  Ford Pinto, 32
  Gosport War Memorial Hospital, 14, 15, 274
  Grenfell Tower fire, 11, 12, 27–45, 264, 270
  Hudson River landing, 268
  Kaohsiung propene explosions, 11, 12
  Nestle, 32
  Nord Stream pipelines, 188, 197, 199, 200, 208
  Quebec Bridge collapse, 165
  San Bruno pipeline rupture, 270
  Three Mile Island, 32
  Titan submersible, 14, 15, 115–137, 264, 274
  Tuskegee Syphilis Study, 17

Distance
cultural, 21, 170, 262–265
geographic, 11, 30, 55, 94, 262–265
hierarchical, 262–265
temporal, 11, 29, 30, 55, 94, 170, 262–265
Diversity, 12, 18, 88, 167, 172–174, 179, 180, 217, 263
Due diligence, 11, 29, 44, 45
Duty, *see* Obligation

E
Economic interests, conflict, 2, 3, 5, 8, 11, 12, 19, 40, 41, 43, 44, 117, 120, 121, 123, 220, 238, 246, 263, 264, 270, 272, 273
Emergency response, 75–78, 80, 84–86, 268
Emotion, 16, 103, 146, 152, 155, 159, 167, 168, 177, 180
Engineering, definition, 71
Epistemic injustice, 14
Epistemic technology, 20
Epistemic trust, 146, 147, 260, 261, 270, 271, 274, 275
Equity, 88, 168, 172–174, 177, 178, 265
Ethical framework, 7, 12, 15, 92, 230, 249
Ethical licensing, 19, 216–220, 225–232, 272
Ethical principles, 6, 12, 19, 88, 101, 104, 105, 108, 111, 165, 178, 253, 260, 272
Ethics, definition, 4–7
Ethics education, 7, 12, 72, 73, 86
Expertise, professional, 2, 4, 10, 31, 37, 40, 62, 73, 115, 118, 132, 166, 190, 218, 239, 241, 246, 269, 271, 273

F
Families, *see* Patients and families

G
Gene editing, 19, 20
Geopolitics, 187–209
Gift of failure, 131
Golden thread, 29
Government policy, 8, 17, 69, 70, 72, 84, 98, 159, 188–193, 195, 199–201, 205, 215, 216, 218–220, 238, 244, 252
Guidelines, 69, 87, 96, 97, 99, 104, 166–168, 177, 180, 260, 264, 265, 272, 273

H
Hallucinations (AI), 261, 268, 276
Hazard pay, 19, 237, 248, 253
Health, 18, 19, 105, 106, 222, 224, 259
Healthcare, 8, 15, 16, 100, 101, 104, 106–109, 262, 264, 265, 275
Hermeneutical injustice, 143, 150
High reliability organizing, 115, 266, 267
Hippocratic Oath, 16–17

I
Identity, professional, 62, 165, 166, 172, 177, 189, 197, 248
Ignorance, 15, 70, 78, 81–86, 171
Inclusion, 18, 88, 172, 174, 178, 200, 206, 239, 251, 253, 254
Industrial development, 12, 70–74
Injustice, 107, 108, 168
Innovation, 120–124, 130–132, 221–225, 232

Insider, organizational, 14, 17, 117, 132–136, 191, 273
Iron ring, 17, 20, 163–180

J
Judgment, *see* Decision making, judgment

K
Kipling, Rudyard, 163, 165, 171, 173, 178
Klara and the Sun, 215
Knockout animals, 225–227, 230
Knowledge
  about risk, 194, 195
  authoritative, 156
  claims, 155
  collective, 29
  divergent, 167
  embodied, 261
  engineering, 74, 166
  giver of, 146
  lack of, 80, 81, 86, 157, 225
  medical, 15, 16, 152
  monopoly, 151
  operational, 125
  personal, 146
  production, 142, 155, 266
  professional, 15, 72, 87, 117, 145, 176, 194, 239, 260, 261, 264, 266, 271, 273
  public, 81, 83–85
  safety, 2, 3, 136
  scientific, 103, 107, 110, 222, 223
  technical, 39, 40, 246
  undervaluing, 159

L
Labor relations, 238, 246, 248, 250
Labor unions, 18, 19, 189, 196, 198, 200, 201, 204, 206, 242, 243, 245–247, 249, 251–254
Legal endogeneity, 50, 52, 62, 63
Legal liability, 70, 80, 119, 152
Legislation, *see* Regulation
Lying at work, 31, 32, 40, 41, 134, 136

M
Macro-ethics, 5, 17, 72, 73, 86, 178, 180
Many hands, 2, 30, 117, 262
Marine engineering, 121, 129
Medical, 8, 16, 17, 20, 96, 214, 219, 223, 225–227, 231, 232, 245, 260, 265, 269, 273
Methods
  disaster cases, 21, 28, 75, 142–143
  discourse analysis, 169–170
  document analysis, 21, 28, 117, 142–143, 188, 243
  interviews, 21, 55–56, 198–199, 220–221
  mixed methods, 220–221
  netnography, 21, 169
  research ethics, 169
  social media data, 169
Micro-ethics, 5, 17, 18, 86, 166, 180
Mill, John Stuart, 54
Moral disengagement, 11, 29, 32, 42, 45
Moral muteness, 10, 11, 29, 43, 45, 164, 169
Museum, 12, 94, 97–99

N

Nader, Ralph, 69
NATO, 188, 191, 193
Negligence, 28, 43, 70, 78, 80, 81, 84, 116, 151, 152, 154
Networks, professional, 9, 168, 179
NGOs, 85, 87
Normalization of deviance, 74
Norwegian Ocean Industry Authority, 188, 198, 199, 203
Nuclear power generation, 54, 269

O

Obligation, 6, 16, 92, 111, 151, 165, 171, 173, 177, 206, 208, 240, 245, 253, 276
Offshore oil and gas, 14, 18, 64, 129, 187–209, 263
Organizational context, 2–4, 8, 13, 14, 29, 32, 40, 41, 45, 115, 117, 119, 131–133, 135, 136, 151, 157, 158, 168, 246, 248, 263, 264, 270–275
Outsider, organizational, 14, 17, 117, 120, 132–136, 261, 273
Over-reliance on AI, 261, 264, 269–271, 275

P

Palliative care, 148–155, 160
Patents, 19, 272
Patients, 7, 14, 16, 265, 267, 275
Patients and families, 14
Petroleum industry, *see* Offshore oil and gas
Pipelines, 11, 12, 49–65, 267, 270, 271
Precarity of work, 264
Pride, professional, 173
Profession, definition, 4
Professionals
  building industry, 11
  developers, 262–265, 269, 273
  doctors, 6, 9, 14, 16, 142, 145–147, 151, 153, 155, 157, 158, 239, 241, 259, 269, 270, 275
  educators, 12, 73, 91, 93, 109, 110, 159, 173
  engineers, 11, 14, 17, 18, 49–65, 115–137, 163–180, 187–209, 239, 241, 259, 262, 264, 270, 272, 273
  lawyers, 62
  maritime, 14, 125
  nurses, 6, 14, 241, 268
  occupational safety and health, 238, 239, 245, 251, 252, 254
  pathologists, 274
  pilots, 1–4, 9, 115, 129, 267, 268, 275
  police, 13, 14, 141, 142, 152–156
  radiologists, 269
  regulators, 18, 56, 63–65, 135, 141
  researchers, 12, 159, 173, 264, 265
  risk consultants, 11, 49–65, 270
  scientists, 19, 213–233, 259, 265, 273
  white collar, 248, 251, 252, 254
Public discourse, 69, 75, 164, 260
Public inquiries
  Grenfell Tower Inquiry, 28
  purpose, 143
  The Report of the Gosport Independent Panel, 141–143
  Titan Marine Board of Investigation, 117
Public trust, 49, 92, 117, 189

R
Rare events, 21, 261, 263, 265–269, 271, 276
Reasonably practicable, *see* As low as reasonably practicable (ALARP)
Regulation, 4, 8, 11, 15–20, 35–37, 39, 41, 42, 44, 50, 55, 61, 62, 64, 80, 82, 85, 96–97, 99, 123, 124, 126, 136, 137, 164, 168, 190, 195, 196, 208, 214, 215, 217–220, 230, 233, 265, 266, 272, 273, 275, 276
Respect, 6, 12, 20, 97, 99–105, 109–111, 151, 230
Responsibility
  to act, 119
  collective, 18, 29, 71, 93, 166, 168, 175–177, 179, 205, 208, 262, 276
  defining, 177
  direct, 124
  drift, 207, 209
  forward looking, 30
  individual, 17, 71, 131, 166, 175, 176
  for learning, 159, 160
  legal, 80–81
  limits, 164
  managerial, 206
  moral, 9, 206, 249
  ontological, 249
  for outcomes, 21, 271
  for past harms, 171
  to profession, 165
  professional, 14, 73, 116, 123, 125, 132, 223, 269
  role and, 125, 127, 165
  for safety, 8, 15, 126, 205
  for security, 188, 205
  shifting to others, 42, 64
  social, 10, 73, 230
  of unions, 206
Retaliation, 14, 117, 126
Reverence, 12, 101, 102, 105, 108–111
Rights, 5, 13, 83, 93, 96, 105, 189, 207, 216, 218, 227, 231, 232, 240, 242, 244, 245, 247, 253
Risk, 10, 11, 19, 49–65, 70, 81–87, 92, 93, 106–108, 117, 118, 126, 131, 136, 167, 177, 187–209, 214, 220, 222–224, 227, 232, 261, 267, 270–273
Risk monetization, 19
Risk society, 74
Risk work, 64
Rituals, professional, 17, 163–180

S
Sabotage, 188, 191–193, 196, 197, 199, 208
Safety, 8, 14, 15, 19, 21
  patient, 13, 14, 17, 267
  public, 2, 3, 10, 11, 18, 49, 50, 55, 56, 58, 62, 64, 82, 84, 86, 167, 215, 222, 223, 264, 267, 270, 271
  system, 1–3, 14, 20, 115–137, 187–209, 261, 267
  workers, 18, 19, 50, 260
Safety imagination, 29, 30, 45, 118
Safety voice, 119
Scientific progress, 229–231
Search and rescue, *see* Emergency response
Security, 5, 18, 187–209, 263, 274
Self-driving cars, 260
Silenced, *see* Speaking up
Social embeddedness, 14, 117, 132–134, 136
Social history, 70–75

Speaking up, 2, 3, 13–15, 21, 106, 116–120, 124–130, 132–134, 137, 207, 265, 271–274, 276
Standards, 28, 42, 49–65, 80, 86, 87, 96–98, 110, 121, 122, 125, 127, 129, 131, 214, 217, 242, 249–254, 265, 272, 273
Sterile seeds, 224
Synthetic biology, 215, 220, 221, 225, 227, 228, 230, 260, 272

T
Technoscience, definition, 4
Teleology, 5
Testing and certification, 2, 3, 33–40, 42, 43, 121–123, 125, 127–129, 131, 271
Threat environment, 189, 208, 274
Tobacco, 224, 225
Tourism, 10
Tradition, 13, 18, 21, 102, 103, 157, 158, 170–172, 177, 179
Tripartite model, 18, 189, 195–206, 208, 209

U
Unconscious bias, 106
Utilitarianism, 5

V
Values
  differing, 3, 5, 7, 9, 10, 12, 13, 15, 19, 20, 132, 261, 272
  professional, 6, 7, 11, 16–18, 20, 21, 40, 49, 62, 64, 73, 117–120, 130–137, 168, 172, 174, 180, 189, 195–197, 205, 206, 208, 216, 218, 249
Virtue ethics, 6

W
Weak signals, 118, 201, 207, 266
Whistleblowing, 2, 4, 13–15, 17, 116, 119, 126, 133–137, 157, 160, 261, 271–274
Workarounds, 19, 228, 229, 231

The manufacturer's authorised representative in the EU is Springer Nature Customer Service Centre GmbH, Europaplatz 3, 69115 Heidelberg, Germany. If you have any concerns regarding our products, please contact ProductSafety@springernature.com

Printed and bound by CPI Group (UK) Ltd, Croydon, CR0 4YY
17/09/2026
02218454-0011